EAST TENNESSEE
CANTILEVER BARNS

EAST TENNESSEE CANTILEVER BARNS

Marian Moffett and

Lawrence Wodehouse

The University of Tennessee Press

Knoxville

Library of Congress Cataloging in Publication Data

Moffett, Marian.

    East Tennessee cantilever barns / Marian Moffett and Lawrence
Wodehouse. —1st ed.

      p.   cm.

    Includes bibliographical references and index.

    ISBN 0-87049-798-7 (cloth: alk. paper)

1. Cantilever barns—Tennessee—Blount County—Design and construction.
2. Cantilever barns—Tennessee—Sevier County—Design and construction.
3. Cantilever barns—design and construction.
I . Wodehouse, Lawrence.   II. Title.
TH4930.M65   1993
690' .8922' 09768885—dc20   92-42651
                                       CIP

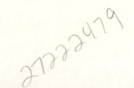

To our parents
Flo, Walter, Raiford,
and in memory of Ronald

CONTENTS

FIGURES

Preface   xi

Introduction   xiii

1.   Cantilever Barns Described   1

2.   Origins of the Cantilever Barn   17

3.   The Builders   27

4.   Regional Factors   47

5.   The Barns of Middle Creek   53

6.   The Barns as Vernacular Expression   73

    Appendix A. Agricultural Graphs   79

    Appendix B. Barn Locations and Dimensions   85

    Notes   125

    Bibliography   133

    Index   139

1.   Two-Crib Double-Cantilever Barn at Tipton Homeplace, Cades Cove (Blount 8)   xiv

2.   Construction of Typical Cantilever Barn (Sevier 29)   xv

3.   Typical Crib Corner with Half-Dovetail Corner Timbering and Corner Foundation Stone   3

4.   Typical Crib Door with Wooden Hinges and Latch (Blount 98)   4

5.   Barn with Taper on Both Primary and Secondary Cantilevers (Sevier 29)   5

6.   Partially Demolished Barn with Long Horizontal Roof Purlin that Supported Lightweight Rafters (Sevier 158)   6

7.   Two Roof Framing Systems   6

8.   Barn Loft with Modified Queen Post Roof Structure (Sevier 29)   7

9.   Sketches and Plans of Five Major Types of Cantilever Barns   8

10.   Distribution of Tennessee Cantilever Barns, by Type   9

11.   Two-Crib Double-Cantilever Barn at Tipton Homeplace, Cades Cove (Blount 8)   9

12. McMahan Barn, Richardson Cove, Sevier County
(Sevier 14)   10

13. Interior of Sharp Barn, Boyd's Creek, Sevier County
(Sevier 74)   10

14. Crib with Tapering Cantilever (Blount 23)   11

15. Robertson Barn, Roberts Schoolhouse Road, Sevier County
(Sevier 28)   12

16. Sharp Barn, Cedar Bluff, Sevier County (Sevier 15)   12

17. Sketches and Plans of Single-Crib Double-Cantilever Sharp
Family Barns, Cedar Bluff, Sevier County   13

18. Four-Crib Double-Cantilever McCampbell Barn,
Lawson Crossroad, Blount County (Blount 9)   13

19. Comparative Dates of Known Barn Builders
in Blount County   14

20. Comparative Dates of Known Barn Builders in Sevier County
15

21. Church of the Nativity, from Peredki, Russia   19

22. Swiss Log Barn with Cantilevered Forebay at Saas, Prätigau,
Canton Graubünden, Switzerland   20

23. Cantilevered Forebay and Stair to Loft   20

24. Conjectural Evolution of Crib Barn Plans   22

25. Boyer Barn, Good Hope Community, Cocke County   23

26. Russian Blockhouse, from Bratsk, Siberia   24

27. Fort Marr Blockhouse at Benton, Polk County   24

28. Swaggerty Blockhouse, east of Parrottsville,
Cocke County   25

29. Late Nineteenth-Century Sketch of
Anatolian Blockhouse   25

30. Caudill Barn, Knott County, Kentucky   26

31. Barger Barn, from Eagle Rock, Botetourt County, Virginia   26

32. Comparative Agricultural Statistics for Blount
and Sevier Counties, 1860-1930   31

33. Sketches and Plans of Andes Family Barns   32

34. Sketches and Plans of Barns near Catlettsburg,
Sevier County   34

35. Sketches and Plans of Sevier County Barns
Constructed before 1860   36

36. Sketches and Plans of Barns along County Road 2486,
Sevier County   39

37. Sharp Barn on Cedar Bluff Road (Sevier 11)   39

38. Barn Builders in Matthew Tarwater Family   41

39. Tarwater Barn on Gists Creek, Sevier County (Sevier 9)   42

40. Sketches and Plans of Tarwater Barns along Gists Creek,
Sevier County   43

41. Sketches and Plans of Ingle Family Barns, Ingle Hollow,
Sevier County   44

42. Ingle Barn on Ingle Hollow Road, Sevier County
(Sevier 29)   45

43. Loft Interior with Continuous Eave and Gable End Vents
in McMahan Barn at Richardson Cove (Sevier 14)   48

44. Double-Height Log Crib (Johnson 1)   50

45. Aerial View of Middle Creek, 1936   56

46. Comparative Plans of Middle Creek Barns   59

47. Family Connections among Barn Builders
Living along Middle Creek Road   61

48. Log Cribs and Cantilevers of Trotter Barn
on Middle Creek Road (Sevier 5)   63

49. Double-Height Log Crib of Trotter Barn (Sevier 19)   66

50. Family Connections among Barn Builders Living
along Roberts Schoolhouse Road   67

51. Construction of Seaton Barn at Seaton Springs (Sevier 23)   68

52. Seaton Barn at Seaton Springs (Sevier 23)   70

53. Two-Crib Double-Cantilever Barn Masked by
    Additions and Enclosures (Sevier 112)   74

54. Cantilever Barns within Category of Crib Barns   77

A1. Agricultural Census Data for
    John Marshall Farm, 1860-80   80

A2. Agricultural Census Data for
    Robert Marshall Farm, 1860-80   80

A3. Agricultural Census Data for
    William W. Webb Farm, 1860-80   81

A4. Agricultural Census Data for
    Pryor L. Duggan Farm, 1860-80   81

A5. Agricultural Census Data for
    John Andes Farm, 1860-80   82

A6. Agricultural Census Data for
    Richard Reagan Farm, 1860-80   82

A7. Agricultural Census Data for
    Matthew Tarwater Farm, 1860-80   83

A8. Agricultural Census Data for
    Tarwater Family Farms, 1880   83

MAPS

1. Tennessee Counties, with Location of Cantilever Barns   xvi

2. Regions Where Barns Were Observed, 1966   2

3. Regions Where Barns Were Found, 1984 to 1992   2

4. Sevier County, with Earliest Surviving Barns   38

5. Map of Middle Creek Community, Sevier County,
   with Surviving Barns   55

6. Middle Creek in 1972 United States Geological Survey   57

7. Middle Creek Community, with Surviving Barns   58

8. Original Land Grants in Middle Creek Community   62

B1. State of Tennessee   116

B2. Barns in Bradley and Meigs Counties   117

B3. Barns in South Blount and West Sevier Counties   118

B4. Barns in South Sevier and South Cocke Counties   119

B5. Barns in South Knox, North Blount,
    and North Sevier Counties   120

B6. Barns in North Sevier, North Cocke,
    and Jefferson Counties   121

B7. Barns in Greene, Washington, and Unicoi Counties   122

B8. Barns in Johnson and Carter Counties   123

TABLES

1. Agricultural Census Statistics for Barn Builders near
   Catlettsburg, Sevier County   33

2. Agricultural Census Data for Fourth Civil District
   of Sevier County, 1880   54

3. Agricultural Census Statistics for Barn Builders
   along Middle Creek Road   64

4. Agricultural Census Statistics for Barn Builders
   along Roberts Schoolhouse Road   69

B1. Information Tables for All Barns Surveyed   86–115

PREFACE

This study grew out of a shared enthusiasm for the vernacular architecture of East Tennessee. As architectural historians teaching at the University of Tennessee, we decided early in 1984 to study cantilever barns. The specimens preserved at Cades Cove in the Great Smoky Mountains National Park were known to us, and we had heard from various sources that there were other barns still standing on farms in the region.

To contact the people we felt most likely to have knowledge of a large number of East Tennessee barns, we wrote to the Agricultural Extension Service leaders in all the Appalachian counties of the state, enclosing a sketch of a cantilever barn and asking for their assistance in locating examples within their counties. We also wrote to the Tennessee Historical Commission for information available from county building inventories. Although relatively few of the mountain counties have been inventoried, Steve Rogers of the commission provided us with barn locations in Meigs and Unicoi counties.

In responding to our inquiry, the Agricultural Extension Leader in Sevier County, Joe Woods, surprised us by asserting that there were about fifty cantilever barns in his county. With the cooperation of John Fox, a reporter for the *Mountain Press News-Record*, several articles about our project appeared in the Sevier County newspaper, inviting readers to call the Extension office if they owned or knew of cantilever barns. The assistance of Mr. Woods and the publicity generated by the newspaper articles ensured hospitable reception at many farms. After preliminary trips to the county, it was apparent that there were more barns in Sevier County than we had first imagined.

We therefore planned a methodical field study and began searching Sevier County with weekly field trips in the fall through spring months. In seeking out those farms where the owners had reported they had a cantilever barn, we came across other cantilever barns as well, and soon we were making a more or less systematic traverse of all the paved and unpaved roads in the county, stopping at all promising barns. What had begun as a project to last several months grew into a multiyear undertaking. Mr. Woods's estimate of fifty barns in Sevier County, which we originally thought to be an impossibly high number, was exceeded within the first year, and we ultimately located 183 barns there, the greatest concentration in any single county.

The Tennessee Historical Commission's survey of Blount County directed by John Morgan was completed in 1984, and those records revealed another 104 cantilever barns located in that county. Throughout the whole of East Tennessee, we have located 316 cantilever barns, which form the basis for this publication.

Additional leads for barn locations have come from a small exhibition and catalog prepared late in 1984 which has circulated to community museums within the region. Information provided by persons responding to newspaper articles or attending lectures presented in conjunction with the exhibition have guided us to barns that we might otherwise not have found. Students in courses we teach have undertaken barn-related research and contributed to our knowledge; this was especially true of Charlotte Burdette, a Sevier County resident who located additional barns and introduced us to the county historian, Beulah Duggan Linn, a veritable treasure-trove of historical information relating to the families in the county. Mrs. Linn's assistance has been invaluable.

Correspondence with scholars studying barns elsewhere has proven helpful. We should particularly mention in this regard Robert Ensminger and his work on the forebay barns of Pennsylvania. We wish to thank Douglas Swaim of the North Carolina Department of Cultural Resources for sharing information about building inventories in the western counties of North Carolina; Lonnie Williams of the United States Department of Agriculture and Harry Williams of the College of Agriculture, University of Tennessee, for providing background on termites and their habitats; Newt Odum, retired farm manager of the Agricultural Experiment Station, University of Tennessee, for informing us about early twentieth-century farming practices in East Tennessee; Harold Core of the College of Agriculture for identifying wood species used in the barns; James Hilty of the College of Agriculture for sharing his collection of antique farm implements; the late Frank Woods of the Department of Forestry, Fisheries, and Wildlife at the University of Tennessee and Betsy Groton of the Division of Forestry, Tennessee Valley Authority, for their assistance with dendrochronology; Tom Ladd of the University of Tennessee College of Business Administration for generously analyzing statistical information on the barns; Charles E. Martin of Alice Lloyd College for information on log barns in Kentucky; LeRoy G. Schultz of West Virginia State University for confirming the absence of cantilever barns in his state; and Cratis Williams of Appalachian State University for background information on southern highland material culture. Colleagues who read the manuscript and provided valuable criticism include Bernard L. Herman of the University of Delaware; Terry G. Jordan of the University of Texas; Allen G. Noble of the University of Akron; and John B. Rehder, John William Rudd, and William Bruce Wheeler, all of the University of Tennessee. Through their efforts, the finished product contains fewer errors; all those that remain belong to the authors.

INTRODUCTION

Visitors to Cades Cove, a restored late nineteenth-century community in the Great Smoky Mountains National Park, drive along an eleven-mile loop road circumscribing the meadowlands of the cove. Near the end of the loop, at the Tipton homeplace, the road turns sharply and descends a small hill, from the top of which a singular barn may be seen in the grassland beyond. Its upper floor looms larger than the lower story, making the building seem top-heavy, and its overhanging profile is quite unlike the simple cabins, outbuildings, mills, and churches that dot the remainder of the cove (figure 1).

For many people in East Tennessee, the Tipton barn is probably the most prominent example of a cantilever barn, an unusual form peculiar to this region. Its base story is composed of two log cribs erected on flat foundation stones in a manner similar to the cabins and other outbuildings constructed in the cove and elsewhere. Above the ground story, however, the similarity to other log buildings ceases. The topmost crib logs extend out to the barn's side ends to become the *primary cantilevers,* and on these rest a series of *secondary cantilevers,* which project front to back, establishing the width of the barn. These secondary cantilevers support the loft floor and become the base for a heavy timber loft frame which completes the barn and unites the separate structures of the two cribs. *Cantilever* is the structural name for the characteristic overhanging hewn beams, and *cantilever barn* is the term used to describe the building type (figure 2).

Even though the Tipton barn is a reconstruction and not original to Cades Cove, having been moved here when the cove was reconstructed by the Park Service as a late nineteenth-century mountain community, it seems completely at home in the gently rolling farmland of its present setting. The barn symbolizes the independent, self-sufficient farms established on the hilly land of East Tennessee in the nineteenth century, and it responds to very practical needs. A cantilever barn the size of that at the Tipton homeplace could accommodate livestock of the average farmstead—a pair of cows and a pair of horses—in the cribs, while the loft provided protected storage for hay, cornstalk fodder, and seed. Under the generous overhang, wagons, sleds, and farm implements could be stored, dry and ready

Figure 1. *A two-crib double-cantilever barn at the Tipton homeplace, Cades Cove, Great Smoky Mountains National Park (Blount 8). This barn, a 1968 reconstruction, was moved into the Park from Cataloochee, North Carolina. Like all well-sited barns, it is placed on a slight rise in the landscape so that rainwater drains away from the building.*

for use. The barn's form was well suited to the rainy and humid climate of the southern Appalachians, acting like a giant umbrella to shelter the cribs from frequent rains; and the open breezeway between the cribs promoted air circulation in the loft to minimize the buildup of damp air. To construct such a barn required no skills or specialized tools beyond those required to erect a log cabin, although considerable initial ingenuity was reflected in the barn's unusual design.

The most compelling feature of cantilever barns is their distinctive silhouette. Within the context of the common architecture of East Tennessee, they stand out as original vernacular forms, dramatic and unexpectedly elegant, particularly when their rough-hewn material and very basic construction techniques are considered. Although many house and barn forms in Tennessee reflect broader trends in the South or southern highlands generally, these notable elements in the rural landscape are unique to East Tennessee and thus deserving of closer study as a distinctive building type.

Field research has led to the identification of 316 examples still standing in the late twentieth century, the vast majority concentrated in Blount and Sevier counties (map 1). For those who wish to understand these buildings, the form of the barns and their geographic distribution raise a series of questions concerning origins of the type and the cultural context in which they were constructed. What were the possible sources for cantilever construction? What accounts for the barns' distinctive features? Why did the particular cantilever form appear here and nowhere else? How were these buildings connected to the larger context of log construction generally and barn design in particular? Who built cantilever barns? What role did these barns play in the local agricultural economy?

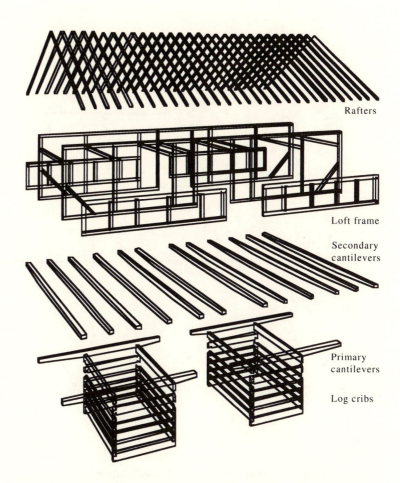

Rafters

Loft frame

Secondary cantilevers

Primary cantilevers

Log cribs

Figure 2. *Exploded diagram explaining the construction of a typical cantilever barn (Sevier 29). The primary cantilevers in this example also extend into the central space, a feature observed on roughly one-third of the barns studied.*

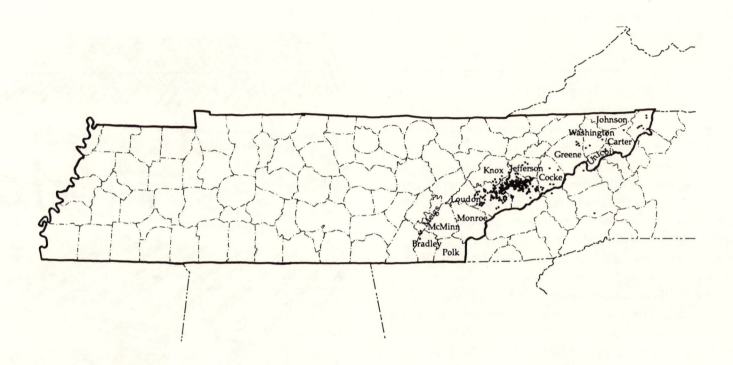

Map 1. *State of Tennessee. Dots indicate the location of cantilever barns, including the three found in North Carolina.*

As an attempt to answer these and related questions, this book sets forth two major theses. First, cantilever barns are seen as a local invention, synthesizing known construction techniques and familiar principles into a new form that enjoyed regional popularity during the late nineteenth century. This thesis is advanced to explain the development and variation observed in barn forms, and it is closely related to the second major theme, which maintains that particular regional characteristics—including relative cultural and economic isolation, the persistence of self-sufficient agricultural economy, and close family connections—account for the construction of so many cantilever barns within such a limited area. The twin threads of invention and tradition will be used to tie together this account of East Tennessee's cantilever barns.

CHAPTER 1

CANTILEVER BARNS DESCRIBED

> In the general area of the Blue Ridge, and particularly the
> Great Smokies of North Carolina and Tennessee, the log
> double-crib barn . . . is found with a large frame loft over-
> hanging in front and back or on all sides by means of the
> cantilever principle.
>
> Henry Glassie, 1965

Henry Glassie was the first investigator to describe cantilever barns. His studies in the 1960s were among the earliest explorations of vernacular buildings of the southern mountains, and his published reports included a map illustrating the range of cantilever barns which served as the initial reference for our field studies (map 2). According to his research, cantilever barns were concentrated in the mountains of western North Carolina and eastern Tennessee, with lesser numbers in the mountains of eastern Kentucky and along the Virginia–West Virginia border. Although Glassie's inquiry indicated approximately equal numbers of cantilever barns found in North Carolina and Tennessee, the actual distribution found in our study indicates that their geographic range is far more limited, confined almost exclusively to Tennessee (map 3). Some of this discrepancy in field studies may be attributed to the passage of time. A quarter of a century separates these two studies, and many barns may simply have disappeared during that interval. Our colleague in West Virginia, LeRoy Schultz, has never succeeded in locating any cantilever barns in his state, so the northern extension of the range shown on Glassie's map remains unexplained.

Virtually all cantilever barns we have located are in East Tennessee, concentrated heavily in two counties, Blount and Sevier.[1] The exceptions are few, and we have not included them

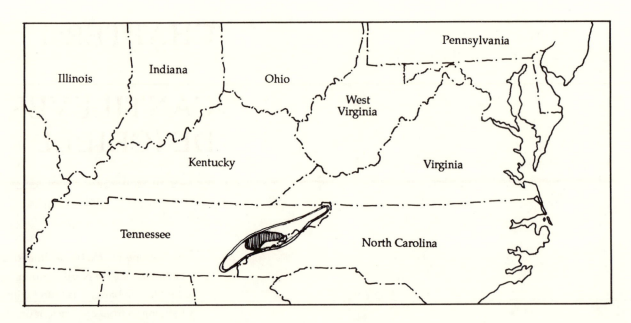

Map 2. *Henry Glassie's map, indicating regions where cantilever barns were observed, based on studies published in 1966. The shaded area contained the highest concentration of barns, while outlying districts had fewer barns. Redrawn with permission.*

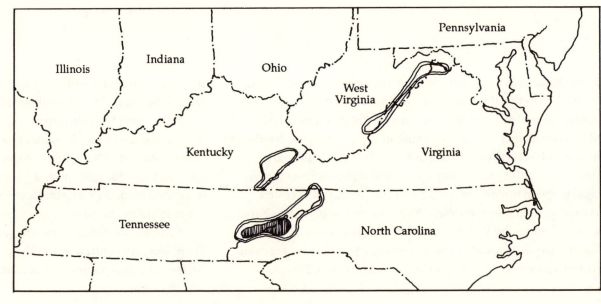

Map 3. *Regions where cantilever barns were found, based on studies undertaken from 1984 to 1992. Blount and Sevier counties account for the overwhelming majority of cantilever barns in Tennessee.*

in this study except as comparison cases. In North Carolina, three cantilever barns have been inventoried in the mountain counties. There are several cantilever barns varying significantly from Tennessee examples in Botetourt County, Virginia.[2] One or two log barns with cantilevers may remain in Kentucky,[3] although we have anecdotal evidence that there once were more. Georgia, likewise, may have had more in the past, although only one standing example is known.[4] As none of the out-of-state barns compare in number, size, or craftsmanship to the best East Tennessee barns, we have concluded that the tendency to apply cantilevers to log cribs is a development unique to East Tennessee, most specifically to Blount and Sevier counties. Blount County has 104 such barns, representing 33 percent of the state total, while Sevier County has 183, or 58 percent of the cantilever barns found in Tennessee. (Summaries of our field records, together with county maps showing barn locations, are presented in Appendix B.) Within both counties, cantilever barns also constitute the overwhelming majority of all log barns. A study by John Morgan and Ashby Lynch found that more than 90 percent of the log barns in Blount County had overhangs,[5] and a similar statistic would be applicable in Sevier County as well.[6]

Characteristics of Cantilever Barns

Cantilever barns share certain characteristics in addition to the log crib base and cantilevered frame upper story. All the barns employ flat stones set on the ground at the corners of the cribs for foundations (figure 3). While this may seem insubstantial, it is similar to the foundation treatment used in log cabins in the region, and it afforded protection from soil dampness while allowing uneven ground to be leveled sufficiently to build the horizontal log courses or rounds. Log construction and the accompanying stone corner foundations also enabled barns to be

Figure 3. *Typical crib corner, showing half-dovetail corner timbering and corner foundation stone. Additional boards have been nailed to the right side wall where original logs were cut out to make an animal feed rack.*

dismantled and moved, as some certainly have been, or to disappear without a trace, as several barns have done since this study began. Without a more permanent foundation, it is impossible to identify the sites of vanished barns.

The cribs were constructed of horizontal logs which usually were ax-hewn to a rectangular shape before being set in position. Where round logs were used, the bark was often left on the log, and the timber was smaller in size. From four to eight log courses (rounds) were laid to form the crib wall, with five being the most common number. The first crib logs ran parallel to the ridge beam to provide the lowest possible threshold sill on the side of the crib where doors would be placed. Corner timbering was most frequently done with the half-dovetail notch, which was used in 208 of 310 surviving cribs. V-notching was found in fifty-eight barns, while square or half notches were employed on thirty-five examples. Nine barns employed a mixture of two notch types, most often a combination of V - and half-dovetail notches. Barns within Sevier County were most likely to employ half-dovetail notching, and V-notching was more commonly found in the extreme northeastern counties of Tennessee.[7] Regardless of the notch type used, the crib logs do not meet to form a solid wall, and the space between logs was never chinked. Doors to the cribs were cut after construction, and the doorway was braced with vertical planks on each side, which were then pegged or nailed into the sawed log ends (figure 4). All details, including door hinges and latches, were crafted in wood. In most barns, the crib doors open to one long side, but in fifteen examples, the cribs open to the central runway, and in one barn, the crib doors open to the barn's ends.

Most cribs seem to have been used originally as shelters for animals, and many still are. Their unchinked construction suits the mild East Tennessee climate, in which year-round pasturing of livestock is possible, rendering the complete enclosure of animals in cribs unnecessary. Cribs were built without floors,

Figure 4. *Typical crib door with wooden hinges and latch (Blount 98).*

and only the largest ones had internal divisions. A feed trough formed out of a hollowed half log and supported on the wall logs was frequently placed in one or both cribs. In a few cases, there are continuous sills extending across the center runway, indicating

Figure 5. *Barn having a pronounced taper on both primary and secondary cantilevers (Sevier 29). Note also that the primary cantilevers extend into the center space. This barn is now removed from the county.*

that there might have been a floor laid there at one time to be used for threshing grain. Wheat was a major crop in East Tennessee during the nineteenth century, although little is currently culti-vated.[8] Only six barns now preserve a threshing floor, but we were told by some barn owners that they had removed sills in their barns so that wheeled vehicles could drive into the central space.

The timbers that form the cantilever beams are substantial— generally from thirty to forty feet long, with cross-sections as mas-sive as ten inches wide by eighteen inches deep. Straight-growing yellow pine is the species most frequently used. Lighter in weight than comparable hardwoods, the pine works more easily yet ages to a rocklike hardness which makes it difficult to nail. In fully 70 percent of the barns, the secondary cantilevers taper at their ends, providing a graceful profile while at the same time reducing dead load and reflecting the diminution of live load at their extremities (figure 5). One is tempted to attribute the cantilever's taper to the

builder's sense of proportion, for it seems unlikely that the con-
structors knew that this was the correct structural shape for cantilever beams. Some beams, in fact, taper so much that they
terminate in a cross-section that is wider than it is tall, reducing
the effective stiffness of the beam. Secondary cantilevers are
distributed visually along the primary cantilever, so that the on-
center dimensions are not consistent. Most commonly, there are
secondaries positioned over the corners of the log cribs, where
the greatest structural strength is concentrated, and secondaries
are often spaced more closely along the cantilevered end of the
primary beam than they are over the crib walls.

The typical cantilever barn loft is not log but framed with
heavy hewn timbers, and all joints are mortise and tenon secured
with wooden pegs. The major uprights align over the corners of
the cribs below, with the end frame resting on the extremities of
the primary cantilevers. These vertical members support a heavy
purlin extending the length of the barn (figure 6). Shorter uprights
at the ends of the secondary cantilevers support an eave beam, and
light-weight rafters rest on the eave beams and purlins, meeting at
the peak of the roof without a ridge beam. Two framing systems
have been observed: one, which is somewhat related to a queen
post truss, has a horizontal tie beam connecting the major uprights
widthwise, and the other, corresponding to a simple post-and-
lintel on which the purlins rest, has no transverse connection
across the barn (figures 7 and 8). (In many barns, the queen post tie
was later cut out to facilitate installation of a mechanical hay
loader.) Both framing systems generally lack significant diagonal
bracing normally found in heavy timber construction, although
knee braces between horizontal and vertical members are commonly observed. Diagonal braces between the loft floor and the
wall end frames are also seen in some barns. Farmers told of wind
damage to barn lofts when an unusually severe storm struck a

Figure 6. *Partially demolished barn revealing the long horizontal roof
purlin which supported the lightweight rafters (Sevier 158). The
original wooden roof shingles are also visible on this collapsing barn.*

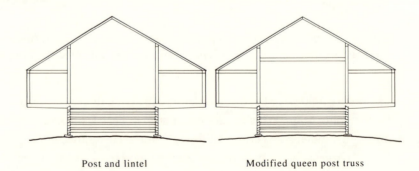

Post and lintel   Modified queen post truss

Figure 7. *Comparative diagrams of the two roof framing systems
observed in cantilever barns. Post-and-lintel structures have relatively
little diagonal bracing between members.*

Figure 8. *Barn loft showing modified queen post roof structure (Sevier 29). Notice the diagonal struts bracing the end frames to the plate or sill resting atop the primary cantilever. This is the barn illustrated in figures 2 and 5.*

farm, and we documented one badly windblown barn which was subsequently dismantled.

On the exterior, the loft wall was generally finished with horizontal lapped siding, cladding that in all cases was produced in a sawmill. The barns were thus built of both hand-hewn and machined wood. The siding was almost never painted. A foot or two of wall just underneath the eaves was left unboarded or covered with widely spaced boards to form a continuous vent. This eave vent usually continued on the end elevations, following the inverted V shape of the gable roof. Originally, all the barns were covered with a wooden shingle roof, known locally as a board roof.

Five Types of Cantilever Barns and Their Variations

The most common cantilever barns found in this study are what we have termed the double-cantilever type with paired log cribs. The word *double* refers to the two sets of cantilever beams which extend both front to back and side to side from the base cribs. In East Tennessee, cantilever construction was also applied to other crib barn types, both single cribs and four-crib plans, and not all cantilever barns were doubly cantilevered. We have identified five types of cantilever barns, some having minor variations, and a small category of unclassifiable irregular forms (figures 9 and 10).

Two-crib double-cantilever barns are the type represented by the Tipton barn in Cades Cove (figure 11). Just over half of the barns surveyed (161 of 316) belong to this classification, and thus it can be considered the dominant type, representing the largest, most handsomely crafted, and (probably) the earliest of all cantilever barns.

Two-crib double-cantilever barns are characterized by having two log cribs separated by a central runway. The top logs of each crib extend out 8 to 10 feet to the barn's ends, becoming the primary cantilevers on which a series of secondary cantilevers, extending from front to back of the barn, are supported. On these beams is laid the loft floor, which in most cases does not continue across the central space. This space is left open to permit a full hay wagon to be pulled between the cribs and unloaded into the loft. A square opening with mitered corners is left in the loft wall across the central space to accommodate high loads.

The McMahan barn in Richardson Cove (Sevier 14) is the largest example found of the two-crib double-cantilever barn

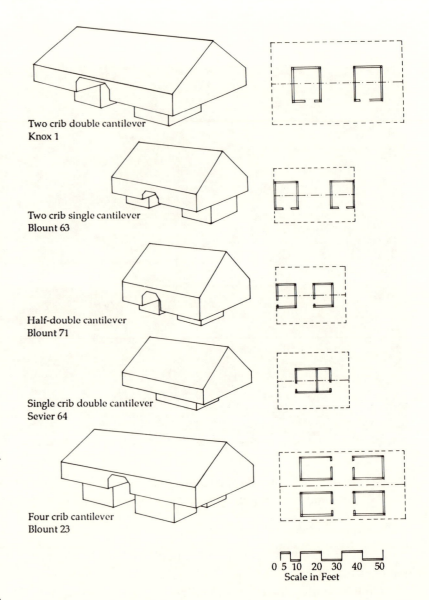

Two crib double cantilever
Knox 1

Two crib single cantilever
Blount 63

Half-double cantilever
Blount 71

Single crib double cantilever
Sevier 64

Four crib cantilever
Blount 23

0 5 10   20   30   40   50
Scale in Feet

Figure 9. *Perspective sketches and comparative plans of the five major types of cantilever barns. The plans of these barns are drawn to the same scale.*

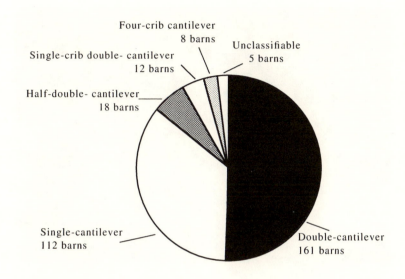

Figure 10. *Pie chart showing the distribution of Tennessee cantilever barns by type.*

Figure 11. *Two-crib double-cantilever barn at the Tipton homeplace, Cades Cove, Great Smoky Mountains National Park (Blount 8). The vertical siding is not typical of barns studied, but the wooden shake roof was original to all barns.*

(figure 12). Built about 1880 by Thomas DeArnold Wilson McMahan (1849-1921), one of the most prosperous farmers in the district, the barn has an overall length of 81 feet 11 inches and a depth of 38 feet. The cribs measure 24 feet by 18 feet, and each is divided transversely into two pens. The loft cantilevers 10 feet on all sides of the cribs, and the central runway is nearly 14 feet wide.

In the Boyd's Creek community is another large two-crib double-cantilever barn, this one built by John Sharp (Sevier 74) (figure 13). The barn's overall dimensions are 69 feet 10 inches by 36 feet 1 inch, with cribs measuring 13 feet 10 inches by 20 feet 1 inch. An unusually wide central runway measuring 24 feet 6 inches separates the cribs, and the twelve heavy secondary

cantilever beams are among the largest found, measuring 10 inches by 16 inches in the center and tapering to a 10-inch by 9-inch cross-section at their ends. The barn at the Tipton homeplace in Cades Cove (Blount 8) is more typical of the size found in two-crib double-cantilever barns (figure 11). Its overall dimensions are 45 feet by 29 feet 4 inches, while the cribs are nearly square, 11 feet 11 inches wide and 11 feet 9 inches deep, and set across a central runway 10 feet wide.

The two-crib double-cantilever form was found with at least four significant variations. In thirty-two of the barns studied, the primary cantilevers also extend a short distance (1 to 5 feet) into the central space (figure 14). This balances the load on the beams

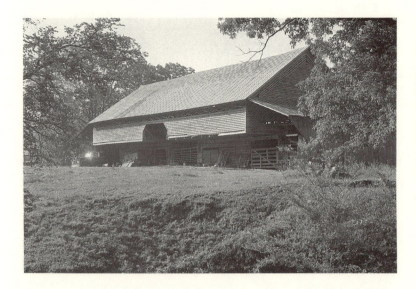

Figure 12. *McMahan barn, Richardson Cove, Sevier County (Sevier 14). This barn, with an overall length of nearly eighty-two feet, was the largest cantilever barn found in this study. Its two cribs are each divided into two pens. As with many barns, there are additions that mask part of the cantilever form.*

Figure 13. *Interior of the Sharp barn, Boyd's Creek, Sevier County (Sevier 74). The unusually wide central space is defined by sill beams and enclosed by boarding. External siding on the loft no longer exists.*

more reasonably while increasing the area of the loft floor above. In four of the two-crib double-cantilever barns (Blount 34, 35, 36, and Sevier 34), the primary and secondary cantilevers are in reversed position; that is, the primary cantilevers extend from front to back of the barn while the secondary cantilevers are placed out to the barn's end walls. The advantages of such a juxtaposition are not readily apparent. In two other barns (Blount 57 and Sevier 183) there are three primary cantilevers built on each crib, the extra beam running across the middle of the crib. By providing extra support through the center of the loft, this design would increase the bearing capacity of the loft floor, particularly when the barn crib was especially deep. One barn (Sevier 112) reversed the posi-

tion of the cribs and cantilevers, so what usually was the side cantilever overhung the central space instead, creating an exceptionally wide—33-foot 8-inch—runway between the cribs (see figure 53).

Two-crib single-cantilever barns are the next most common in occurrence. These barns resemble the classic form in all respects except that there are no primary cantilevers (and thus no overhang at the sides or ends of the barn), only secondary cantilevers placed across the crib and extending from front to back. Since there is only one set of cantilevers, this is termed a single-cantilever barn. Obviously, such a form does not create as large a loft space as in barns that are doubly cantilevered, and the overall size of single-cantilever barns is smaller. In general, the quality of construction

Figure 14. *Crib with tapering cantilever extending into central space (Blount 23).*

was lower and the structural timbers in these barns were smaller than in double-cantilevered barns . Of the barns surveyed, 112, or 35 percent, were of this type.

Half-double-cantilever barns combine aspects of the first two types in equal amounts. These are termed two-crib half-double-cantilever barns, because one crib is doubly cantilevered and the other one has only single cantilevers (figure 15). Eighteen such barns were found, representing only 6 percent of the total, but the type occurs often enough that it does not seem to be accidental. No particular reason for this hybrid form can be deduced, although in some cases topographic obstacles may have inspired the variation. The single- and double-cantilevered cribs do not appear to have had different functions, as for a livestock pen and a corn crib, so the rationale behind half-double-cantilever barns remains elusive.

Figure 15. *Robertson barn, Roberts Schoolhouse Road, Sevier County (Sevier 28). This is a half-double-cantilever barn with the doubly cantilevered crib located on the left-hand side.*

Figure 16. *Sharp barn, Cedar Bluff, Sevier County (Sevier 15). A single-crib double-cantilever barn, it has large openings in the loft wall to permit loading.*

Single-crib double-cantilever barns have a single crib and doubly cantilevered loft. In the twelve barns of this type that were found, the crib is considerably larger than those found in double-crib barns and typically is divided into two pens of unequal size (figure 16). Since there is no central space, loading the loft had to be accomplished through an opening or openings in the barn's loft wall. All but three of these barns are clustered in the eastern half of Sevier County, suggesting that the design might have been a narrowly localized variation. Three were built by the Sharp family within one-half mile of one another on Cedar Bluff Road (Sevier 11, 12, and 15); barns 12 and 15 have similar proportions and details (figure 17). Their cribs measure approximately 26 by 16 feet;

the primary cantilevers project 10 feet out to the sides; and the secondary cantilevers are only slightly shorter.

Four-crib cantilever barns represent the application of cantilever principles to the four-crib barn, a plan type that Glassie has suggested might be original to southeastern Tennessee. His research found symmetrical four-crib barns with square cribs built as far south as Alabama and Mississippi, but the older plan with rectangular cribs is typical primarily of the Smokies.[9] All eight of the four-crib cantilever barns found were in Blount and Sevier counties, most of them having rectangular cribs and runways of approximately equal width. The cantilevers in both directions are usually minor, less than 4 feet. The McCampbell barn at

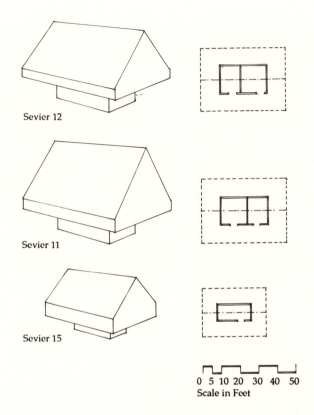

Sevier 12

Sevier 11

Sevier 15

0 5 10 20 30 40 50
Scale in Feet

Figure 17. *Perspective sketches and comparative plans of the single-crib double-cantilever barns built by the Sharp family, Cedar Bluff, Sevier County.*

Lawson Crossroad (Blount 9), dated pre-1900 by the Tennessee Historical Commission survey, illustrates the type (figure 18). Its cribs measure 14 by 12 feet, while the runway through the center of the barn is 12 feet 2 inches wide and the transverse runway measures 11 feet in width. The side cantilevers extend out 8 feet, but the overhang projecting to the front and back of

Figure 18. *McCampbell barn, Lawson Crossroad, Blount County (Blount 9). A four-crib double-cantilever barn having slight projections to the front and back. Additional construction disguises the end cantilever, while a corn crib attached to the right side extends the barn.*

the barn is only 15 inches in length. The barn's overall dimensions in plan are 56 feet 3 inches by 37 feet 11 inches.

The Dates of Cantilever Barns

Dating these barns has been more difficult and problematic than obtaining dimensions. Reliable tax records do not exist, and property records do not include descriptions of the improvements on a given piece of land. It has been possible to conduct dendrochronological analysis on only a handful of the total number of surviving barns, but that too has proven inconclusive, since dated softwood specimens from local trees of the nineteenth century apparently are not available.[10]

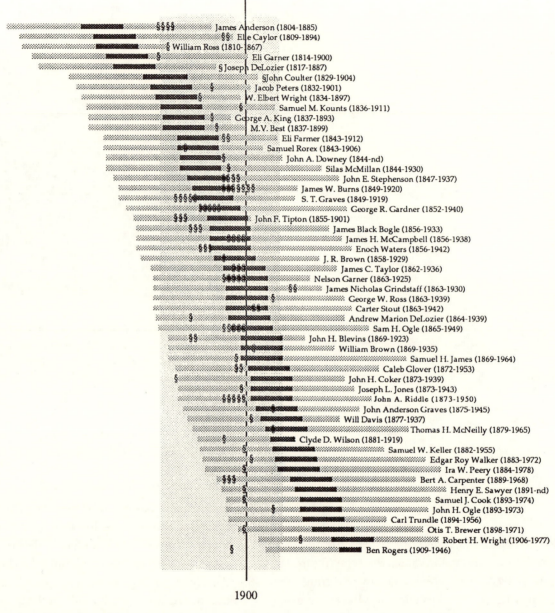

§§§§ James Anderson (1804-1885)
§§ Elle Caylor (1809-1894)
§ William Ross (1810-1867)
§ Eli Garner (1814-1900)
§ Joseph DeLozier (1817-1887)
§John Coulter (1829-1904)
§ Jacob Peters (1832-1901)
§ W. Elbert Wright (1834-1897)
§ Samuel M. Kounts (1836-1911)
§ George A. King (1837-1893)
§ M.V. Best (1837-1899)
§§ Eli Farmer (1843-1912)
§ Samuel Rorex (1843-1906)
§ John A. Downey (1844-nd)
§ Silas McMillan (1844-1930)
§§§ John E. Stephenson (1847-1937)
§§§§ James W. Burns (1849-1920)
§§§§ S. T. Graves (1849-1919)
§§§§ George R. Gardner (1852-1940)
§§§ John F. Tipton (1855-1901)
§§§ James Black Bogle (1856-1933)
§ James H. McCampbell (1856-1938)
§§§ Enoch Waters (1856-1942)
§ J. R. Brown (1858-1929)
§ James C. Taylor (1862-1936)
§§§§ Nelson Garner (1863-1925)
§§ James Nicholas Grindstaff (1863-1930)
§ George W. Ross (1863-1939)
§ Carter Stout (1863-1942)
§ Andrew Marion DeLozier (1864-1939)
§§§§ Sam H. Ogle (1865-1949)
§§ John H. Blevins (1869-1923)
§ William Brown (1869-1935)
§ Samuel H. James (1869-1964)
§§ Caleb Glover (1872-1953)
§ John H. Coker (1873-1939)
§ Joseph L. Jones (1873-1943)
§§§§ John A. Riddle (1873-1950)
§ John Anderson Graves (1875-1945)
§ Will Davis (1877-1937)
Thomas H. McNeilly (1879-1965)
§ Clyde D. Wilson (1881-1919)
§ Samuel W. Keller (1882-1955)
§ Edgar Roy Walker (1883-1972)
§ Ira W. Peery (1884-1978)
§§§ Bert A. Carpenter (1889-1968)
§ Henry E. Sawyer (1891-nd)
§ Samuel J. Cook (1893-1974)
§ John H. Ogle (1893-1973)
§ Carl Trundle (1894-1956)
§ Otis T. Brewer (1898-1971)
§ Robert H. Wright (1906-1977)
§ Ben Rogers (1909-1946)

1900

Figure 19. *Comparative dates of known barn builders in Blount County. The shaded area represents the period between the Civil War and the First World War. Dark portions of the time lines indicate the years when builders were 30 and 45 years old, the most likely period for construction of a barn. The § sign indicates the date assigned to the barn by the Tennessee Historical Commission survey.*

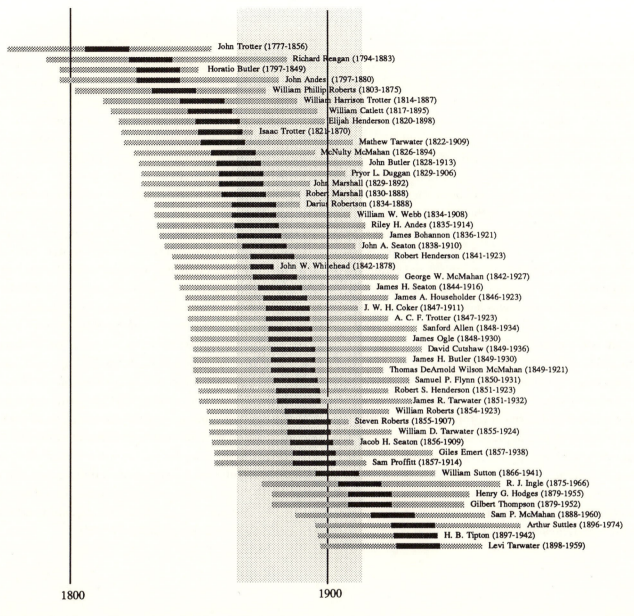

Figure 20. *Comparative dates of known barn builders in Sevier County. The shaded area represents the period between the Civil War and the First World War. Dark portions of the time lines indicate the years when builders were between 30 and 45 years old, the most likely period for construction of a barn.*

From cemetery records and interviews with barn owners, dates have been established for some of the farmers who built the barns. If this dating information may safely be regarded as representing the whole, it appears that cantilever barns were constructed from about 1815 until 1938 (figures 19 and 20).[11] It seems clear that a number of cantilever barns, perhaps 12 percent of those remaining, were constructed prior to the Civil War, and an equal percentage in the period between the war and 1880. Fully 50 percent were built in the two decades between 1880 and 1900, with rapidly decreasing numbers until the 1920s. The most recent confirmed date of a cantilever barn is 1938 for the Suttles barn (Sevier 170), a two-crib single-cantilever barn of small unhewn logs put together with rudimentary skill. During the Great Depression, there was a general return to log building in Tennessee's rural counties, so the Suttles barn may reflect a late revival of log construction using the cantilever principles of cultural tradition.

Cantilever barns were probably not built by organized or professional construction teams, as were log buildings in other places[12] and some midwestern barns of comparable date. Our study has produced no evidence of professional builders, and there are several reasons which might account for this. First, the two-crib barn was becoming an obsolete type by the second half of the nineteenth century. Published plans for barns and outbuildings advocated more specialized accommodation for specific animals or farm activities instead of the general volumes suggested by two-crib designs. "Modern" barns of the late nineteenth century were built in frame, not log.[13] Second, from the data gathered in this study, the types, sizes, features, and proportions of these barns vary considerably even within a small area, providing limited evidence of standard building dimensions; and neither field inspection nor statistical analysis have revealed traits that might constitute a builder's signature.

Of course, the individual farmer would not have built his barn single-handedly or without reference to existing barns in the neighborhood. Local carpenters may well have been employed on both house and barn construction in portions of a given county, especially in the period after the Civil War. The custom of cooperative raisings for houses was still alive early in this century,[14] and this same process was probably used to raise barns as well. Older Sevier County residents have recalled that certain men were once known as "corner men" because of their particular skill at cutting the notches for log construction. Working in pairs, these men would adjust the notches to keep each log course horizontal, ensuring that the structure rose evenly. Neither popular memory nor documentary accounts from the region mention professional log builders in the nineteenth century. It would thus appear that cantilever barns are traditional creations, architectural forms so well understood and generally utilized that their design and construction could be accomplished by anyone who could build a log cabin and master braced frame construction in heavy timber.

CHAPTER 2

ORIGINS OF THE CANTILEVER BARN

The survival of 316 cantilever barns in the region indicates that the type was once very common, and it raises significant questions about the origins of the form. Historic and cultural traditions point to several possible sources for the overall shape and construction of cantilever barns: log construction, which was extremely common along the American frontier; the forebay bank barns of the Pennsylvania Germans, which have cantilevered overhangs on one side; English and German barns, which feature heavy timber framing similar to that seen in the lofts of cantilever barns; both German and Swiss barns, which provide the precedent for two-crib plans; and log frontier forts or blockhouses, which were built with cantilevers over all four sides.

Previous studies of American folk buildings have tended to account for vernacular design features through the process of cultural diffusion, whereby traditions of a given European "cultural hearth" are transposed to the New World, mingled with related ideas, and recreated across the continent. By this reasoning, aspects of Appalachian barns and outbuildings have been traced to prototypes in the German settlements of Pennsylvania, and the pioneering studies of Henry Glassie set cantilever barns as a special case of Pennsylvania barns in the South. While not denying the impact of German two-crib barn plans and the cantilevered

forebay of the great Pennsylvania barn, we believe that the blockhouses which were demonstrably present in the region, having been constructed by the early settlers, may have provided inspiration for the builders of the region's first cantilever barns. The blockhouses' geographic proximity and parallel construction lead us to propose that they are the strongest single influence on the cantilever barns' design. These conclusions can be supported through an examination of the evidence.

Log Construction

The tradition of log building was not indigenous to America, even though the idea of the log cabin is inseparable from images of early American life. Neither are log buildings found in the English, Welsh, Scots, or Irish cultures from which many American colonists came,[1] although techniques of log building had been developed and widely used in Europe during the medieval period, particularly in the heavily forested regions of central and northern Europe, Scandinavia, and western Russia.[2] Horizontal log construction may well have originated in Russia; log grave houses built by nomadic horsemen on the steppes of Siberia have been uncovered and dated to the fifth century B.C.[3] From Russia, the idea of horizontal log construction passed to

Scandinavia, where it replaced upright log or stave construction around 1000 A.D. Many scholars agree that horizontal log construction was introduced to the New World around 1638 with Finns and Swedes who settled in what is now Delaware.[4] From this beginning, the technique spread to Anglo-Saxon and Celtic immigrants and was reinforced and augmented by eighteenth-century immigration from log-building regions of Germany and central Europe. Rather elegant European details were greatly simplified by American settlers, whose primary objective was to erect shelter quickly. Therefore, some details that became very common on this side of the Atlantic, such as half-dovetail corner timbering, were not often found in Europe.[5] The speed with which a log building could be erected, the limited range of tools and techniques needed for its construction, and the large number of trees in the forested areas of North America combined to make log buildings common across the frontier.[6] The technique was particularly persistent in the southern Appalachians.

In the area of American vernacular building in log, as with folk building as a whole, the most satisfactory model to explain eighteenth- and nineteenth-century developments is that of cultural diffusion, which traces American forms and construction techniques back to Old World prototypes. Both C. A. Weslager and, more recently, Terry Jordan and Matti Kaups have demonstrated the European origins of log building elements as they have sought to establish an understanding of American work in log. The pioneering studies of Appalachian buildings conducted by Henry Glassie also substantiated the model of cultural diffusion, showing that the Appalachian log cabin was an amalgam of German and Scots-Irish cultures, both of which contributed to the mountain population.[7]

As was demonstrated in chapter 1, most East Tennessee cantilever barns appear to date from the post-Civil War period after the initial pioneering effort was largely over. When farmers had brought sufficient acreage under cultivation and obtained crop yields sufficient to justify a sizable barn, quite a few of them chose to build a cantilever barn. These barns thus correspond to what Jordan describes as the "second generation of log structures," those buildings more carefully crafted and generally constructed several decades after the period of first settlement.[8] Even when portable and permanent water mills increased the availability of sawed timber, people in rural East Tennessee continued log building traditions. Killebrew's summary study of the entire state in 1874 noted of farm buildings in East Tennessee: "These are generally built of wood. The dwelling-houses are often of plank, but most generally of logs. They are neither handsome, comfortable nor convenient, as compared with the better class of houses."[9] It appears that many of the log barns documented in our study were being built at the same time that some houses and churches in East Tennessee were being constructed in light-weight balloon frame, although log houses certainly continued to be constructed in the southern mountains well into the twentieth century.[10] Log construction and sawed timber were not mutually exclusive. Indeed, the combination of ax-hewn wood in cribs and loft frame with sawmill-produced external siding is typical of cantilever barns, and some of the loft frames are composed of sawed timber as well. Morgan's study of East Tennessee log houses suggests that the desire for new, more up-to-date "stylish" houses, combined with the new technology of sawmills and machine-made nails which facilitated economical construction of balloon frame or box building, contributed to the decline of log house construction.[11] Whether similar attitudes influenced barn construction is difficult to say, but we think not. The barn was a functional element on the farm, whereas the house was more likely to be influenced by fashion and the desire for comfort. In addition, the small quarters of most log cabins would have provided cramped living space at best, while a good-sized cantilever barn would not be outgrown so easily. Cantilever barns persisted in part because they continued to fill a useful role.

Precedent for Cantilever Design

As with the tradition of log construction, the principle of the cantilever is of considerable antiquity. It is related to the corbel, a structural technique widely used in very early brick or stone construction, in which each successive horizontal layer projects slightly beyond the course underneath, so that the structure gradually arches out to enclose space. Masonry corbels are necessarily small to prevent tension cracks in the brick or stone. When constructed in wood, longer projections or cantilevers are possible, since timber's tensile strength is significantly greater than that of stone. The cantilever became part of medieval traditions in timber construction: half-timbered or *fachwerk* houses in England, France, Germany, and Switzerland frequently featured small cantilevers called jetties on their upper floors, in part because this made the heavy timber joints easier to construct while having the further advantage of increasing the floor area. This same technique is also found in farm house-barns in Germany, Switzerland, and Austria, where it is often joined with log construction and used for balconies or lofts.[12] In *The Craft of Log Building,* Hermann Phleps illustrates storehouses, granaries, and farmhouses from Scandinavia through central Europe which feature cantilevers on log construction, often constructed on successive cantilevers stacked into a console-like element.[13] Using this same principle, enormous cantilevers were constructed as early as the sixteenth century to support galleries of log churches in Karelia, once the Finnish homeland but now incorporated into Russia (figure 21).

Figure 21. *Church of the Nativity from Peredki, Russia, now at the open-air museum at Novgorod. Dated before 1539, this church has enormous cantilevers supporting the raised gallery.*

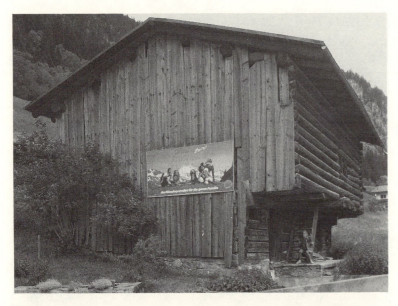

Figure 22. *Swiss log barn with cantilevered forebay at Saas, Prätigau, Canton Graubünden, Switzerland. The barn is built into the hillside so that access to the second-floor loft occurs at grade. Log construction is used throughout, including the side wall covered in vertical boarding.*

Figure 23. *Detail of the barn in figure 22, showing the cantilevered forebay and stair providing loft access.*

Robert Ensminger, a scholar of the German barn in Pennsylvania, has traced the log forebay barn in Europe and found them concentrated in the eastern and central sections of Switzerland, particularly in the valley of Canton Graubünden known as Prätigau (figures 22 and 23).[14] These Swiss forebay barns are entirely constructed of small-diameter round logs for cribs, cantilevers, and lofts. Their design is that of a bank barn, set with the long side against the hill and having a cantilevered forebay on the opposite side. A staircase under the forebay provides connection between levels. Even the roof frame is log: long logs span between the end walls to serve as purlins for the roof.

German Barn Types

Not only were cantilever construction and log techniques found in European architecture, but also the precedent for most American barn types has been traced to European prototypes. In particular, the double-crib barn is traceable to Germany, where it is prevalent in the Rheinland-Palatinate region, the homeland or way-station for many German immigrants to Pennsylvania.[15] Terry Jordan, in a study completed in 1985, has also found double-crib log barns in central Switzerland and eastern Austria.[16] The studies of Ensminger and Jordan have established that the transfer of Germanic barn types to Pennsylvania occurred as settlers from the Old

World modified the buildings they knew to fit conditions in the New World. Glassie's studies of barns in the eastern United States led him to interpret the Pennsylvania barn as an American development.

> Just as the diverse early dwelling traditions of the groups which settled in southeastern Pennsylvania gave way to a new house type less than a century after the colony was founded, so the small early barn types of Britain and the Continent were largely supplanted by the late eighteenth century with the two-level "bank barn." The cattle are stabled in the basement of the Pennsylvania barn and the upper level, used for hay grain storage, is cantilevered over the lower on the barnyard side. This overhang, which generally faces the south, east, or some compass point between, is called an "overshot," "foreshoot," or "forebay"—*der forschoos* or *der forbau* in the Pennsylvania Dutch dialect. The barn is built into the bank or has a ramp—"barnhill"—or bridge built up to the side opposite the forebay, so that the vehicle can be drawn into the floor of the second level. It is built of stone or brick, log or frame with a stone basement.[17]

Crib barns appear to have been among the earliest of American barns to develop, and a chart illustrating their evolution has been drawn by Noble (figure 24),[18] elaborating on the diagram proposed earlier by Kniffen.[19] The most basic barn form, a single log crib, may be constructed to shelter animals or crops, and more elaborate barn forms may be made by combining two or more cribs, with or without common walls, into a single structure. The two-crib double-cantilever barn is constructed from identical log cribs placed side by side to form a central space, with cribs, central space, and a significant amount of perimeter land all sheltered by the second-floor overhanging

loft. Noble's chart shows the Tennessee cantilever barn to be a special case of all double-crib barns, but its evolution from the larger category is not explained. Glassie grouped cantilever barns in the general category of double-crib barns, of which he designated three types; cantilever barns were identified as a subset of type II double-crib barns. As with many of the vernacular buildings he found in the Appalachians, Glassie identified structures built by German settlers in Pennsylvania as the probable source for these barns. Specifically, he noted the similarity of the overhanging loft in the Appalachian barns to the forebay of the great Pennsylvania barn, which is generally built with one long side against a hill and has a cantilevered forebay projecting on the opposite side. Since the great Pennsylvania barn most typically has a base story of stone, not log, and the Appalachian barns are freestanding, not built into a hillside, Glassie observed that a German peasant house, the *Umgebindehaus*, might have provided the formal precedent for the cantilever barn because this medieval peasant house had similar form and materials: a log base and frame upper story.[20] While it is true that the *Umgebindehaus* has the requisite structural systems, the manner in which the two connect is quite different from what has been observed in East Tennessee barns. The *Umgebindehaus* frame appears to be supported on sturdy timber posts which are set forward of the log base, whereas the frame of cantilever barns rests squarely over the corners of the log cribs. Although it is indisputable that German and Swiss cultural influences came down the Shenandoah Valley of Virginia from Pennsylvania into Tennessee, it is difficult to understand how knowledge of the *Umgebindehaus*, a German house type never built in the New World, could have been transmitted to the Scots-Irish who settled predominantly in Appalachia.

Cantilever barns might be seen more plausibly as adaptations of the traditional Pennsylvania bank barn, commonly built in log

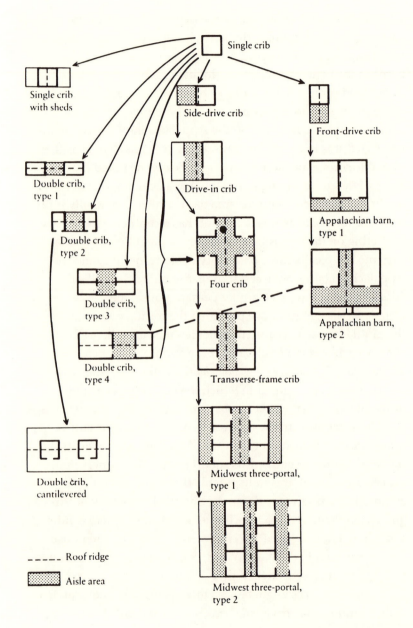

during the eighteenth century but better known for the later versions with masonry base and timber-frame forebay. Several Tennessee barns illustrate the possibility of this transference of ideas, and their survival into the 1980s indicates that there may well have been more at some earlier time. The most striking connection with Pennsylvania tradition was the Boyer barn in the Good Hope community of Cocke County, which was said to date from the early 1800s (figure 25). This bank barn, located within the earliest and largest settlement area of Pennsylvania Germans in East Tennessee,[21] had a handmade brick lower story completed by a heavy timber frame upper floor with a south-facing forebay supported on twenty-six cantilevered beams. The barn measured approximately 47 by 34 feet in plan, not including the 8-foot cantilevered forebay. (Dimensions are approximate because they were scaled from photographs.[22] The barn was destroyed by fire in August 1985 before there was an opportunity to measure the building.) Vertical siding on the forebay does not appear to be original, but the horizontal lapped boarding on the end elevations is typical of the siding found on cantilever barns.

Within Knox County there are four surviving bank barns on the general model of the Pennsylvania barn, one of which retains log construction in its base story. Three of these bank barns have connections to Pennsylvania: two were built between 1870 and 1880 as part of a Mennonite settlement in west Knox County, and the third is also associated with an early German settlement.[23] Although the dates of these barns place them as

Figure 24. *"Conjectural evolution of crib barn plans" by Allen G. Noble. (Figure 1.1 in* Wood, Brick, and Stone: The North American Settlement Landscape, Vol. 2: Barns and Farm Structures. *Reproduced with permission.)*

Figure 25. *Boyer barn, Good Hope Community, Cocke County. Constructed in a German settlement area, with a cantilevered forebay projecting over a handmade brick base story. (Photograph by Gary Hamilton.)*

contemporaries rather than predecessors of cantilever barns in East Tennessee, their presence here establishes additional ties to German settlements in Pennsylvania. They demonstrate that the Boyer barn was not an entirely isolated occurrence.

Blockhouse Construction

One possible inspiration for the four-sides cantilever comes from the design of military blockhouses, which seem by the mid-eighteenth century to have become a nearly universal type,

perhaps as a reflection of commonalities in military instruction and a sharing of defensive practices throughout western Europe. Wooden stockades have an ancient history in Europe, and wooden forts with log watchtowers having doubly cantilevered upper floors were even built across Siberia from the seventeenth to the middle of the nineteenth century (figure 26). Two seventeenth-century watchtowers from Bratsk in Siberia survive in open-air museums,[24] and their form recalls that of frontier blockhouses built in the United States. Blockhouses were an integral part of protective settlement in eighteenth-century Tennessee. Although several appear on early maps, at least six documented examples were constructed in East Tennessee just after the American Revolution to guard major back-country military supply routes against Indian attack.[25] Of the two that survive, Fort Marr in Polk County outside Benton, in the extreme southwest corner of Tennessee, is built entirely of squared logs, and its slight doubly cantilevered overhang is comparable to the Russian example (figure 27).

The Swaggerty blockhouse, built in Cocke County in 1787,[26] most closely matches the fundamental features of a cantilever barn: a log base with half-dovetail notching, and a doubly cantilevered log upper story finished with horizontal lapped siding (figure 28). The similarity of the Swaggerty blockhouse to a drawing made in the late nineteenth century of a log blockhouse in central Anatolia is striking,[27] so much so that one speculates that a type of "international style" in military architecture existed when defensive positions were established in undeveloped areas (figure 29). Jordan proposes a German origin for blockhouses, noting that the word *blockhaus* in German means log house, but he also suggests an English military connection, perhaps through German soldiers who fought for the British in America. Jordan further notes the similarities between block-

Figure 26. *Russian blockhouse from Bratsk, Siberia, now at Kolomenskoe Park in Moscow. A fragment of the log palisade wall is included against the left side of the blockhouse.*

Figure 27. *Fort Marr blockhouse at Benton, Polk County.*

houses and doubly cantilevered Scandinavian granaries.[28] The use of heavy timbers for construction and the presence of a projecting upper story are generally accepted characteristics of blockhouses, and although few eighteenth-century American blockhouses survive, they were built both as free-standing structures, as in East Tennessee, and as corner bastions of a log palisade enclosing a more substantial fort.

Structures combining log construction, the cantilever prin-

ciple, and a double-crib barn plan within a single building were not unique to East Tennessee. Robert Ensminger has provided information about the Corle barn in Pennsylvania, in which all these characteristics are present, but the particulars of its design are distinctly different from East Tennessee cantilever barns. The Corle barn has double-height round log cribs to which the cantilevered forebays of hewn logs are attached. Thus, there is no loft in frame construction, as is common in Tennessee canti-

Figure 28. *Swaggerty blockhouse located east of Parrottsville, Cocke County.*

Figure 29. *Anatolian blockhouse sketched by German travelers in the late nineteenth century. (From Petersen and von Luschan.)*

lever barns, although the external profile of the Corle barn is close to Tennessee models. From photographs, it appears similar in structure to the Caudill barn in eastern Kentucky, where saddle-notched round logs were used to construct a two-story divided log crib (figure 30). Midway in the height of the barn, three logs project as cantilevers on the long dimension of the barn, but further details of the barn are lost in the building's ruin. From Botetourt County, Virginia, comes the Barger barn,

an intact example built along similar lines (figure 31), now relocated to the Frontier Culture Museum of Virginia at Staunton. Its double-height cribs have a cantilevered forebay front and rear, but the side cantilevers are paired logs which extend to support a shed extension to the roof rather than an upper loft floor. The barn, dated to approximately 1855, also has a threshing floor. Neither the Kentucky nor the Virginia barn appears to be framed as are most East Tennessee cantilever barns.

Figure 30. *Caudill barn, Knott County, Kentucky. In its ruined state, the original form of the barn is not entirely clear. The double-height crib appears to have had a major cantilevered bay on one long side. (Photograph by Charles Martin.)*

Figure 31. *Barger barn from Eagle Rock, Botetourt County, Virginia, now at the Frontier Culture Museum of Virginia at Staunton. The barn features double-height log cribs with front-to-back cantilevers providing a loft floor. Cantilevers extending to the ends of the barn support only the sheltering roof. A wooden threshing floor is set between the cribs.*

We are thus inclined to interpret East Tennessee cantilever barns as local inventions which became widely adopted as traditional forms within a limited geographic range. Although the idea of cantilevered log construction was explored in other locales, nowhere else did it achieve the popularity and varied development observed in this region. We believe these barns to represent one of few original American forms in log construction.

CHAPTER 3

THE BUILDERS

If cantilever barns were local inventions, who were the inventors? What were the backgrounds of the families who built them, and what might account for the localized variations observed among barns? An attempt to answer these questions led us to examine the settlement history of East Tennessee and study the population and agricultural census data for families who built barns. These inquiries supported the proposition that the barns were local inventions because no other explanation seems to fit the evidence gathered.

Barn-building families represented a number of ethnic groups common to the southern highlands—Irish, English, Scots-Irish, and German—and they moved into East Tennessee from Virginia and North Carolina. Only in East Tennessee, however, did they or their descendants build cantilever barns. Therefore, cultural diffusion—the theory most frequently advanced to explain developments in folk culture—does not seem to apply in this particular case. Someone, perhaps from within Sevier County, must have originated the form, and it met with sufficient local enthusiasm that the cantilever principle was applied to single-crib, two-crib, and four-crib types. Family connections, geographical proximity to an existing cantilever

barn, and possibly the social prominence of several early barn builders seem to us to account for both the distribution and small-scale variation in these barns. An overview of the people who built these barns is therefore in order.

Early Settlement in East Tennessee

During the second half of the eighteenth century, East Tennessee gradually gained settlers of European descent. Before the coming of these outsiders, the indigenous Cherokee Indians had permanent towns in the valley of the Little Tennessee River, and they used the upland areas as hunting grounds. They sowed crops in the fertile valleys, which were kept free of large trees through burning; a traveler in 1805 noted "that the spacious meadows in Kentucky and Tennessee owe their birth to some great conflagration that has consumed the forests, and that they are kept up as meadows by the custom that is still practiced of annually setting them on fire."[1]

Around 1750, hunters and explorers established trading contacts with the Cherokee, generally for furs and skins; and in 1756, the British built Fort Loudoun on the Little Tennessee near its junction with the Tellico River to provide protection for English colonists in North and South Carolina. Although by treaty the entire region west of the Appalachians was off-limits to European settlement, colonists from North Carolina, Virginia, and Pennsylvania, stimulated by favorable reports from explorers and encouraged by land speculators, began to move into the northern portions of the region after 1763, when the conclusion of the French and Indian War removed French claims to the territory. Both individuals and groups negotiated privately with the Cherokee for leaseholds or purchases of land. This unauthorized migration continued during the American Revolution, when the land became a part of North Carolina and several towns established in the future state of Tennessee, including Rogersville, Jonesborough, and Greeneville, gained recognition by North Carolina. Regiments composed of settlers contributed to the British defeat at the Battle of King's Mountain (1780) and engaged in numerous skirmishes with the native peoples.

In 1783, North Carolina enacted legislation which threw almost all its Tennessee land up for sale at extremely low prices, to the benefit of land speculators. Only the region east of the Tennessee River and south of the French Broad and Big Pigeon rivers was reserved for the Cherokee, and, as settlers of European descent were already there, immigrant pressure to obtain clear rights to this Indian land increased. Dissatisfaction with North Carolina's government and its management of the territory contributed to the formation in 1784 of the State of Franklin, a short-lived confederation of upper East Tennessee settlers who sought admission as the fourteenth state in the union. The new state was condemned by North Carolina and not recognized by the national government, but nevertheless its officers pursued agreements with the Indians. In June 1785, the Cherokee were compelled by the Treaty of Dumplin Creek to permit Franklinites to settle south of the French Broad as far as the Hawkins line, a surveyed boundary running

west northwest to east southeast just south of present-day Maryville. In November of that same year, however, the Treaty of Hopewell negotiated between commissioners of the national government and the Cherokee reversed the Dumplin Creek agreement, restoring Indian claims to mountain land east and north of Greeneville. Since this placed much of the State of Franklin, including its capital, Jonesborough, on Indian land, the frontier settlers fought the Cherokee, forcing them in August 1786 by the Treaty of Coyatee to cede territorial claims to all land north of the Little Tennessee River. However, the treaty was quickly violated. Dissension within the leadership of Franklin and reconciliation with North Carolina authorities brought about the demise of the State of Franklin, and after North Carolina ratified the United States Constitution in 1789, control of its lands west of the mountains passed to the federal government.

In 1790, Federal jurisdiction over the region was established with the creation of the United States Territory South of the Ohio, which was to provide for orderly settlement and lead to the formation of new states in the land across the mountains. Leadership was entrusted to Governor William Blount, a prominent North Carolina land speculator for whom Blount County was named. In 1791, he and the Cherokee signed the Treaty of the Holston, which opened up land east of the Clinch River and north of the Hawkins Line to white settlement. To a great extent, this treaty legitimized what had been in effect for some time; and, although some Cherokee leaders disputed the authority of the tribal elders to relinquish so much ancestral land, the provisions of this treaty were not reversed. Two later treaties, one in 1798 and another in 1819, terminated remaining Cherokee claims to land in East Tennessee.[2]

The settlers who populated East Tennessee came from various places. Many, particularly those of German and Scots-Irish descent, had moved down the valley of Virginia from settlements in Pennsylvania in search of better land. From Virginia and North Carolina came colonists of English, Scots, French, or Scots-Irish background, some seeking freedom for their religious beliefs and others escaping the corrupt land policies of the lords proprietors and their agents. These land-hungry and independent-minded individuals found in the rugged terrain of East Tennessee the space they sought for homesteads and the freedom to live without significant governmental interference. From southwest Virginia, the migration route followed the Holston River, funneling settlers from Abingdon into the Tennessee Valley. Indian paths and tributary river courses provided access to the smaller valleys and coves, enabling settlers to disperse over uneven land.[3]

The first lands claimed by speculators or settlers were the easily farmed fertile valleys. Here it was possible to establish relatively large farms, which might be farmed by slaves. In more remote areas, however, self-sufficient farming was the rule from the earliest European settlements until well into the twentieth century. As an account of East Tennessee agriculture written in 1874 describes the farms and farmers of the region,

> [t]he most distinguishing characteristic of the average farmer in East Tennessee is the effort which he makes to supply what may be required for his own consumption. He is indeed a great provider of the necessities of life. He is ambitious to live within himself. It is not uncommon on a small farm to see a patch of cotton, which the women of the household work up into cloth; a spot given to tobacco for home consumption; a field of sorghum from which syrup is made for domestic use; a few acres of wheat are raised for flour; corn and oats or hay to feed the stock, which usually consist of a few sheep to supply wool for winter clothes, cows from

which a considerable revenue is derived by the manufacture of butter, and a brood-mare or two from which the farmer rears his mules and horses for farm use. Besides these, an abundance of the standard vegetables, such as cabbage, beans, peas, potatoes and onions, is raised, as well as ducks, chickens, geese, guinea-fowl, peafowls, &c. A few bee-hives, and an apple and peach orchard, are the necessary adjuncts to nine-tenths of the farms in East Tennessee. The most striking fact in the farming operations in that division, is that no money-crop, so-called, is raised. Tobacco, cotton, corn and hay, are all grown in small quantities, not so much for sale as for use. The amount of money realized by the average farmer of East Tennessee is painfully small, and yet the people in no portion of the State live so well, or have their tables so bountifully furnished.[4]

The Agricultural Profile of Cantilever Barn Builders

To understand better the farmers who built cantilever barns, we have traced the names of known barn builders in Blount and Sevier counties through both the population and agricultural census records and verified their birth and death dates from cemetery records (see figures 19 and 20).[5] Since the investigation of barn builders has not been without certain ambiguities and uncertainties, some preliminary caveats might be in order. First, we have taken the builder's names given to us by present barn owners as being correct. Where possible, these have been cross-referenced with other sources, but many names have been accepted because there is no other means of corroboration, and the assumption that these are the actual builders may not necessarily be valid. When locating names in the population and agricultural censuses, we have looked for individuals residing in the correct census district, assuming that the farmer thus located was indeed the builder of a cantilever barn. (No addresses are given in either census.) This process broke down when multiple individuals in a given district had the same name. In 1880, for example, there were two adults named Will Davis in the four-teenth civil district of Blount County. In such cases, it was im-possible to determine which was the one we sought. When both fathers and sons shared the same name, we have relied on prob-able dates to establish responsibility for barn construction. As a result of limited information, our set of barn builders is not nec-essarily a random or representative sample of the whole, being composed solely of individuals we could identify from cemetery records or find in the population census. Barns without any builder's names are obviously excluded, as are barns whose build-ers are known only by surnames common in the county. Of these builders, detailed agricultural information is available only for those who lived at the right time to be enumerated in the 1850 to 1880 agricultural censuses. Some identifiable builders were either too old to be reflected accurately in 1850 or too young to be regis-tered in 1880, the interval in which agricultural census data for individual farms were collected and released. Since 1880, only aggregate farm statistics have been provided by the Bureau of the Census, and they reveal little that is of interest to this study.

The information on builders is also characterized by vagar-ies in the census itself. Statistics gathered vary from decade to decade; census takers sometimes provided initials and some-times wrote out names in full; boundaries of census districts do not remain constant and are often omitted from the tabulation altogether; and names are spelled differently from census to census. For example, the family name *Cutshaw* becomes *Cutshall*, and *Robertson* is sometimes spelled as *Robbison*. One may won-der at the accuracy of agricultural data: why did farms often report one hundred eggs a year, regardless of the number of poultry, and why is annual butter production on many farms also relatively uniform? How reliable are statistics supplied by a farmer who cannot write? The 1880 population census included questions regarding literacy, revealing that some farmers and their adult children or wives were unable to read and/or write. Killebrew, the sometime state commissioner of agriculture, re-marks that the agricultural census data are little to be trusted.[6] Nevertheless, the census is the only primary source for agricul-tural history in nineteenth-century East Tennessee, and we have drawn on it to reconstruct the farm economy in Blount and Sevier counties at the time cantilever barns were being built.

In very general terms, Blount and Sevier counties have com-parable agricultural conditions; their soils, crops, and livestock are similar (figure 32). Blount County farms, however, tended to be larger and more prosperous than those in Sevier County, a trend that continued from 1850 through the first decade of the twentieth century. In both counties, the number of farms be-came larger and their average size and value grew smaller as the turn of the century approached. Thus, a farmer whose hold-ings during this period remained fairly constant improved his standing relative to the average of all farms.

As explained in chapter 1, dating cantilever barns has proven problematic. Where birth dates for builders have been established, we have assumed that barns were built when the farmer was between thirty and forty-five years old. This range accords well with those few cases where we have both a verifi-able construction date and the birth/death dates of the builder. In general, this interval also appears to coincide with the period when, according to census statistics, the farm's crops, livestock, and value approximated the average for that period. In later life, many barn builders rose above average, in part because their farms were stable while averages decreased. Agricultural pro-

Comparative Agricultural Statistics for   ░░░ Blount County
Blount and Sevier counties, 1860–1930   ▬▬▬ Sevier County

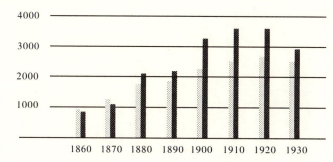

Number of farms

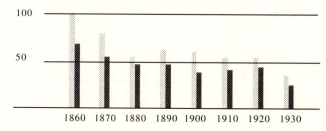

Improved acreage per farm

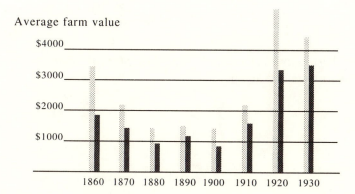

Average farm value

files of three contemporary barn builders—John Marshall (1829-92) of Sevier 61, Robert Marshall (1830-88) of Sevier 137, and William W. Webb (1834-1908) of Sevier 22—demonstrate the general tendency for increased farm prosperity on the part of cantilever barn owners (see graphs in Appendix A).

The case of Pryor L. Duggan (1829-1906) of Sevier 7 also illustrates this situation well, largely because he was the right age to be counted in three of the four detailed agricultural censuses.[7] Duggan's family was large—nine children are enumerated in the census—and a descendent owns the property today. Duggan's father Wilson (1803-75) was born in Sevier County and first appears as a head of household in the 1840 census. Wilson was a schoolteacher, lawyer, and farmer who represented Cocke and Sevier counties in the Tennessee General Assembly from 1843 to 1853 and served again as the Sevier County representative from 1865 to 1867.[8] In 1850, the twenty-one-year-old Pryor was still listed in his father's household in the Third District of Fair Garden, in the northeastern part of the county. In 1852, Pryor married a neighbor, Matilda Houk (1832-1922), and by 1860 he had moved to the southwest sector of the county in the Sixth Civil District (later the Sixteenth) of Walden Creek and had fathered four children. The 1880 census notes that Matilda was unable to write, but Duggan's family was not illiterate; his older brother, Lemuel Duggan, was the Assistant Marshall who enumerated the entire 1850 Agricultural Census for Sevier County. We estimate that Pryor Duggan's barn was built between 1860 and 1870, making it one of two known Civil War-era cantilever barns in the Walden Creek community of Sevier County.

Figure 32. *Comparative agricultural statistics for Blount and Sevier counties, 1860-1930.*

The other early barn in Walden Creek, Sevier 32, was the property of John Andes (1797-1880), who was also included in the 1860-80 agricultural censuses.[9] Because the earliest statistics on the Andes farm come from the period when John Andes was over sixty years old, we cannot accurately envision his farm in the preceding decades. He was listed in the 1830 census with a wife and three female children under the age of ten, and he lived to be counted in 1880. From what one can reconstruct from the census records, Andes and his wife Lettie Murphy appear to have been parents of at least seven children. John, Lettie, and one son are buried at Shiloh Methodist Episcopal Church cemetery. John Andes is interesting on several counts. Both he and his parents were born in Virginia, showing that cantilever barns were not always the products of second- or third-generation farmers in East Tennessee, although his family does conform to the general pattern of settlers moving from Virginia into the mountains of Tennessee.

Two of Andes's sons, Riley H. Andes (1835-1914) and John W. Andes (1838-1919), established adjacent farms near Catlettsburg, just north of Sevierville, where both built cantilever barns (Sevier 18 and 3, respectively) larger than their father's (figure 33).[10] The Andes brothers' farms at Catlettsburg were substantial, justifying their spacious barns. In 1870, both farms were the same size and were equal in value, and they remained closely matched although not identical in 1880 (table 1). Among barn builders, they are unusual for growing fiber (cotton or flax), making cheese, and reporting a harvest of hay and clover seed. John W. Andes served in the Civil War, represented Sevier County in the Tennessee General Assembly from 1889 to 1891, and moved to Knoxville in 1901, where he became a claims and pension attorney.[11]

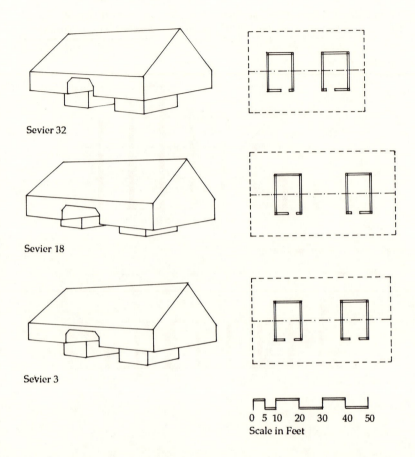

Sevier 32

Sevier 18

Sevier 3

0 5 10   20   30   40   50
Scale in Feet

Figure 33. *Perspective sketches and comparative plans of cantilever barns built by the Andes family.*

Census Year	Improved acreage	Unimproved acreage	Cash value of farm	Value of machinery	Horses	Asses and mules	Milch cows	Working oxen	Other cattle	Sheep	Swine	Value of livestock	Wheat (bushels)	Corn (bushels)	Oats (bushels)	Wool (pounds)	Irish potatoes (bushels)	Sweet potatoes (bushels)	Butter (pounds)	Hay (tons)	Molasses (gallons)	Honey (pounds)	Wood cut (cords)	Other
William Catlett (1817–95) Barn 37																								
1850	325	385	5000		60	5	6				150	4885		2000	200									
1860	175	300	10000	400	4	2	5	2	20		80	1710	200	2000	200		15	15	400		30	300		50 lb cheese; 100 bu peas.
1870	350	480	13000		1	2	5	2	6	43	8	1105	150	1000	150	150	10	2	100	6				Paid $200 in wages.
1880	200	197	4800			1	1					125	400	1000	100				150				35	2 acres apples; 2 bu peas.
Riley H. Andes (1835–1914) Barn 18																								
1870	52	156	3000	300	4		5		8	26	22	990	70	300	275	50	40		200	12				20 lb cheese; 60 lb flax.
1880	88	135	5000	540	2		2		4	26	13	895	95	400	200	70	8	50	250	1.5	36	80	80	50 lb cheese; 2 acres apples.
John W. Andes (1838–1919) Barn 3																								
1870	52	156	3000	300	5		2			14	14	894	84	300	300	30	10	15	200	12		80		40 bales cotton; 22 bu seed.
1880	132	180	5500	660	3	2	7		1	26	10	500	130	500	150	50	10	15	300	2	100	150	50	20 lb cheese; 5 acres apples.
George Washington McMahan (1842–1927) Barn 2																								
1880	264	325	8600	175	3		3		20		42	636	275	400	60			30	270			50	20	5 acres apples; 2 bu peas.

Table 1. *Agricultural Census Statistics for Barn Builders near Catlettsburg, Sevier County.*

The Andes brothers' barns may have been inspired by the proportions of a neighboring barn (Sevier 37), built by William Catlett (1817-95). The dimensions of the Andes brothers' barns are closer to Catlett's barn than to their father's (figure 34). Catlett's barn is located across the Little Pigeon River from the Andes barns, within sight of them. Around 1810, Catlett's father, Benjamin Catlett (1797-1833), opened a tavern in Sevierville at the end of Main Street, and William eventually inherited both the business and his father's slaves. While a farmer, he was also an industrious businessman, continuing the tavern operation and organizing the Bank of Sevierville in 1888, of which he was the first president.[12] The 1850 census lists his occupation as that of a horse trader, although he also maintained an extensive farm.[13] The value placed on his real and personal property make him one of the wealthiest among known builders of cantilever barns. In 1850, his farm (land and buildings) was valued at $5,000, while his livestock including sixty horses was worth $4,885. In 1860, his real estate was valued at $10,000 and his personal property was worth $30,215. The 1870 census indicates that he owned $13,000 in real estate and $7,520 in personal property, a reduction that may partially be explained by the Civil War. By 1880, his animal dealing had declined and Catlett himself was living in the town of Sevierville; however, he was still cultivating over 150 acres in wheat, corn, and oats. Neither the census nor the cemetery records indicate that he ever married.

There is one other cantilever barn in the vicinity of Catlettsburg, Sevier 2, built by George Washington McMahan (1842-1927), and it too approximates the general dimensions established by Catlett's barn. McMahan appears to have inherited the farm of his father Abraham (1800-83), exchanging roles as head of the household between 1870 and 1880. One assumes the barn was built about this time.[14] G. W. McMahan was a second-generation Tennessean whose paternal grandparents were

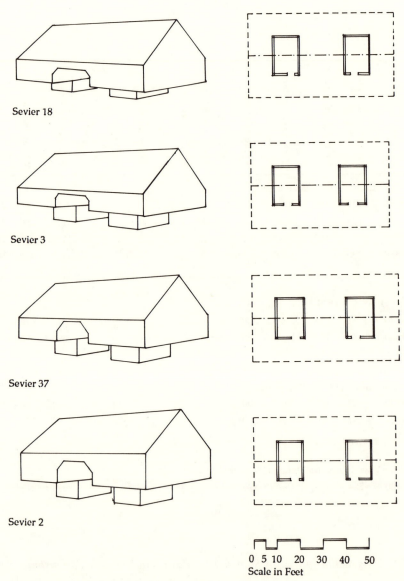

Sevier 18

Sevier 3

Sevier 37

Sevier 2

0 5 10   20   30   40   50
Scale in Feet

Figure 34. *Perspective sketches and comparative plans of cantilever barns near Catlettsburg, Sevier County.*

born in Ireland; his grandfather, James McMahan, held a four-hundred-acre land grant on territory between the west and east forks of the Little Pigeon River, including the present site of the Sevier County Courthouse. Despite this wealth in land, the 1880 census noted that his father Abraham could not write. G. W. and his wife, Marusia V. Henderson (1848-1923), were parents of one daughter, born in 1872, providing them with one of the smallest families among barn builders.[15]

According to the population census of 1880, the Catlettsburg community was an integrated one, with black families living adjacent to the (wealthier) white farmers. While some blacks worked as farm laborers, at least one was identified as a being a blacksmith. The agricultural census reveals that blacks were employed at three of these farms in 1880: R. H. Andes paid $170 for fifty-two weeks labor, John W. Andes paid $200 for the same amount of help, while G. W. McMahan was the most generous employer, paying $150 for twenty-two weeks of outside labor.

Richard Reagan, an Early Barn Builder

Catlett's barn and the original John Andes barn seem to have been seminal barn designs. Were there other influential pre-Civil War cantilever barns? Doubtless there were, but the only ones of which we have specific knowledge are located in Sevier County. Seven presumed barn builders there were born before 1820: John Trotter (1777-1856) of Sevier 5, Richard Reagan (1794-1883) of Sevier 52, Horatio Butler (1797-1849) of Sevier 136, John Andes (1797-1880) of Sevier 32, William Phillip Roberts (1803-75) of Sevier 173, William H. Trotter (1814-87) of Sevier 6, and William Catlett (1817-95) of Sevier 37. Their barns would appear to be verifiably the oldest among surviving cantilever barns. All are double-cantilever types, and all are generously sized and constructed of substantial timber (figure 35).

Four of the seven are located in the Middle Creek community and will be discussed more fully in chapter 5, leaving the barn of Richard Reagan for consideration here.

Richard Reagan settled by 1830 in the western section of Sevier County, in the area of the Tenth Civil District known as Dupont, a name derived from the lower dew point of a local summer resort on a mountain top.[16] Both Reagan and his wife, Phoebe, were born in Virginia; his father was also a native Virginian, while his mother was born in Ireland. This family background is consistent with the general pattern of Scots-Irish settlers moving into Tennessee from Virginia. Reagan's family included at least ten children.[17] Four of them, aged forth-three to fifty-five, were living with their widowed father in 1880, and all were illiterate. The spelling of family names fluctuates over the censuses so much that one suspects that the entire household was lightly educated: the surname is written as Reagan, Rogan, Ragan, and Regan in one decade or another, while Phoebe is variously Phebe or Phebea, only her tombstone in the Reagan-Dupont cemetery providing the common spelling. Arithmetical skills were also lacking. Assuming that his grave marker birth date is correct, Richard never gave the census taker his correct age, and by 1880 both he and his children were only approximately accurate about their own ages.

Reagan's agricultural record is among the most complete available, even though we do not have documentation of his early years in farming (see appendix A). His was an unusual farm within the county in employing mules rather than horses as draft animals. He maintained relatively large herds of sheep and swine and planted a larger-than-average range of crops. In addition to the usual corn and wheat, Reagan reported raising peas and beans, both Irish and sweet potatoes, apples, and tobacco, while also producing honey and beeswax. For all four reporting years covered by the agricultural census, he main-

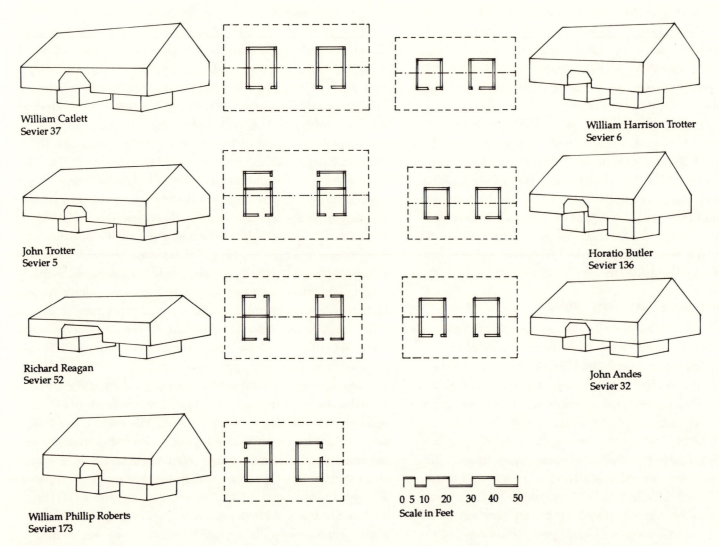

William Catlett
Sevier 37

William Harrison Trotter
Sevier 6

John Trotter
Sevier 5

Horatio Butler
Sevier 136

Richard Reagan
Sevier 52

John Andes
Sevier 32

William Phillip Roberts
Sevier 173

0 5 10   20   30   40   50
Scale in Feet

Figure 35. *Perspective sketches and comparative plans of Sevier County cantilever barns possibly constructed before 1860.*

tained this diversity, unlike other farmers who shifted over time to raising a more limited range of crops. An uncommon detail in his barn may have stemmed from the need to accommodate more animals: each log crib is divided into two compartments by a transverse wall. Given his herd of sixteen to forty sheep, the extra pens would have been particularly useful at lambing time, when additional protection against both the weather and predators should be provided to ewes and lambs. Whether his barn influenced others in the immediate vicinity is unclear, as there are neither obvious physical similarities to neighboring barns nor any records of family ties to barn builders elsewhere.

Crop information from the 1880 agricultural census shows that Reagan's wheat yield of 6.8 bushels per acre was close to the county average (5.1 bushels per acre) and slightly above that of his near neighbor, James Pickens (5 bushels per acre). For corn, Reagan's record of 5 bushels per acre was substantially below both the county average of 17.8 bushels per acre and the yield of Pickens's farm (15 bushels per acre). One can only speculate whether this reflected poor farming practice by Reagan who was now an elderly man, albeit one aided by two grown sons, or whether he was planting on poor or exhausted soil. Agricultural data on both crop size and acreage were collected for the first time in the 1880 census, so comparative information from previous decades is not available.

The first builder of a cantilever barn cannot be identified with certainty, although the distribution of pre-Civil War barns would indicate that it was someone in the Middle Creek community, where four of the seven are located (map 4). The scattered locations of the remaining three, however, show that the idea, wherever it originated, spread rather widely. Accounting for that diffusion has led us to examine the relationships both of barn forms and barn-building families.

Local Groupings of Similar Barns

Finding a collection of similar barns in close proximity to one another, as in the quartet at Catlettsburg, is not an exceptional occurrence. The clustering of barns with common features happened so often that one might conclude that very local variations on the basic cantilever barn types were likely to be copied in neighboring barns. Several examples might be cited to illustrate this point. In Sevier County along County Road 2486, there are three double-cantilever barns (Sevier 46, 47, and 48) set within a half-mile of one another, all having quite similar dimensions, continuous primary beams spanning the central space, and insubstantial side overhangs. Two also have curious steep staircases built along the end of one crib (figure 36). To some degree, these stairs resemble those found on Swiss forebay bank barns, although the location at the barn's side rather than at the front was not typical of Swiss practice. Nowhere else in East Tennessee were similar stairs found, the access to barn lofts invariably being made by ladders attached to the walls of cribs. The builders of the barns along Road 2486 are unknown, as are the reasons for including the stairs in this unusual trio.

Another peculiar grouping involves three single-crib double-cantilever barns erected in Sevier County along Cedar Bluff Road (Sevier 11, 12, and 15) by members of the Sharp family.[18] (See figure 17.) The single-crib double-cantilever barn itself is a relatively rare type, with only eleven having been found in East Tennessee. Finding three of these concentrated in one small area would seem to indicate a quite localized preference for the form. As the Sharp barns exhibit substantially different proportions, their outward appearance is varied, but the underlying structure is common to all (figure 37).

At a somewhat larger scale, the four cantilever barns found in the northeast section of Greene County are related in having

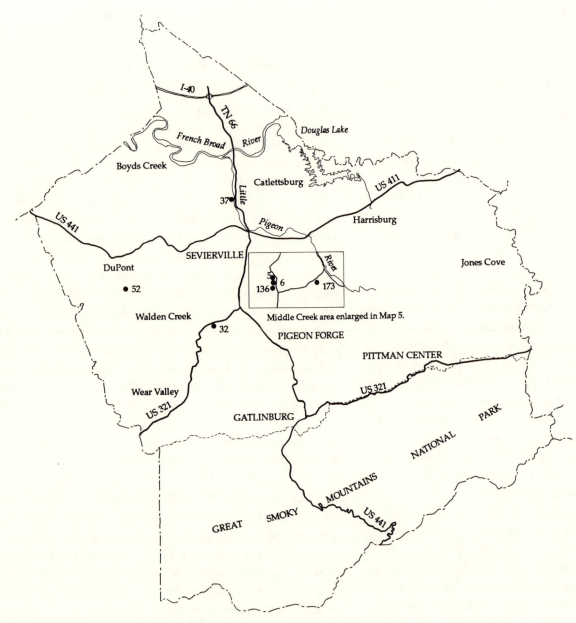

Map 4. *Sevier County, showing locations of the earliest surviving cantilever barns.*

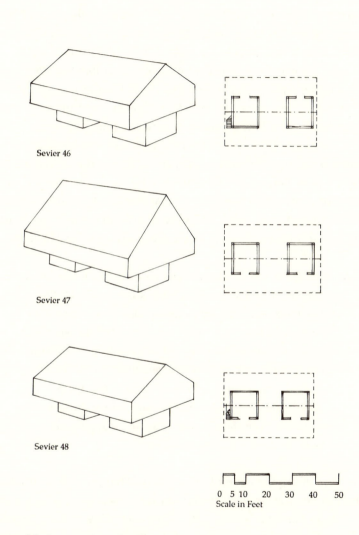

Sevier 46

Sevier 47

Sevier 48

0 5 10   20   30   40   50
Scale in Feet

Figure 36. *Perspective sketches and comparative plans of cantilever barns along County Road 2486 in Sevier County.*

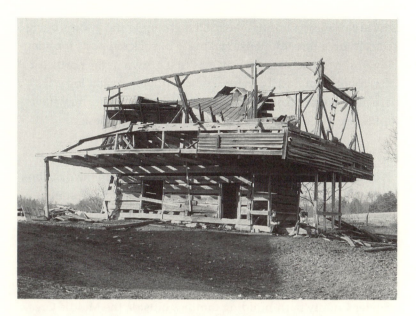

Figure 37. *Sharp barn on Cedar Bluff Road (Sevier 11). The loft of this barn was seriously damaged in a violent windstorm, and the barn has since been demolished.*

two-story log cribs rather than the usual second-floor loft over a single-story crib. Why they were built this way remains somewhat of a mystery. Nelle Starr, the present owner of Greene 4, recalls that grain, especially wheat, was stored in one crib where the bins are still in place, while the other undoubtedly held animals, as evidenced by a log trough. A plank floor had been installed above head height, which, together with a loft across the central space, created three upper-level compartments for hay or fodder. Access to the cribs' upper story was gained through large openings cut in the log wall. Other two-story log cribs in the area had no flooring, the full height up to the purlins being filled with a three-dimensional maze of stick supports for hanging burley tobacco. As burley was first grown in the northeastern counties of East Tennessee where double-height log cribs are most common, it is possible that tobacco culture may have influenced this local variation in cantilever barn design.

Families of Barn Builders

Not only does geographic proximity play a role in the distribution of barn designs, but so too does the connective tissue of family ties. This has already been observed in the case of John Andes. Although the father's farm was not particularly near to those of his sons, one suspects that their inclination to build cantilever barns was in part influenced by the cantilever barn on their father's farm. Our research has uncovered two additional Sevier County barn-building families outside the Middle Creek community—the Tarwaters, who settled along Gists Creek just west of Sevierville, and the Ingle family on nearby Ingle Hollow Road.

The progenitor of the Tarwater line was Matthew Tarwater (1822-1909), who moved from Knox County into Sevier County in 1857 and thus first appeared in the 1860 census with his wife Sarah Rule (1828-1884) and seven children (figure 38).[19] Matthew and Sarah were Tennessee natives, but Sarah's parents were Virginians. Of their nine children, at least five of the seven sons established farms in the vicinity of Matthew's tract. The 1900 population census lists William Dowell Tarwater as family 289, Matthew Tarwater as family 291, Henry Rule Tarwater as family 292, and Matthew Nelson Tarwater as family 293, while showing that Adam Harmon Tarwater, the youngest child, was living with his wife and three children in the household of Matthew senior. Some early members of the family are buried in the Tarwater Cemetery established by the father at the edge of his property, while others are interred at nearby Pleasant Hill Cemetery. Both James Rogers Tarwater (1851-1932) and William Dowell Tarwater (1855-1924) built cantilever barns across from their father's place on present-day US 411, and the youngest son of William, Levi Tarwater (1898-1959), also constructed a cantilever barn, completing a three-generation family tradition.

The farm of the elder Matthew is most completely documented through census records, and it is also numbered among the Century Farms identified in 1986 by the Tennessee Department of Agriculture (see graph in Appendix A). Matthew and Sarah Tarwater acquired 127 acres of land in Sevier County in 1857, and their holdings would eventually grow to 317 acres.

> A religious family, the Tarwater men assisted in the construction of the Pleasant Hill Methodist Church. On Saturdays, the women cooked all the Sunday meals and before dawn on Sundays, the family completed "all the labor and chores" so that "Sundays were free from labor and distractions." The Civil War, however, shattered the Tarwater's hard-working, pious world. According to family history, Confederate troops robbed the family of food, livestock and money. The "family also suffered much personal abuse" for their Union sympathies.[20]

Assuming the general principle that barns were built while the farmer was between thirty and forty-five years of age, then Matthew Tarwater's barn dates from late 1850s or early 1860s (figure 39). The present owner dates the barn to the mid-1880s, a figure that might be related to the survey of this barn in 1984.[21]

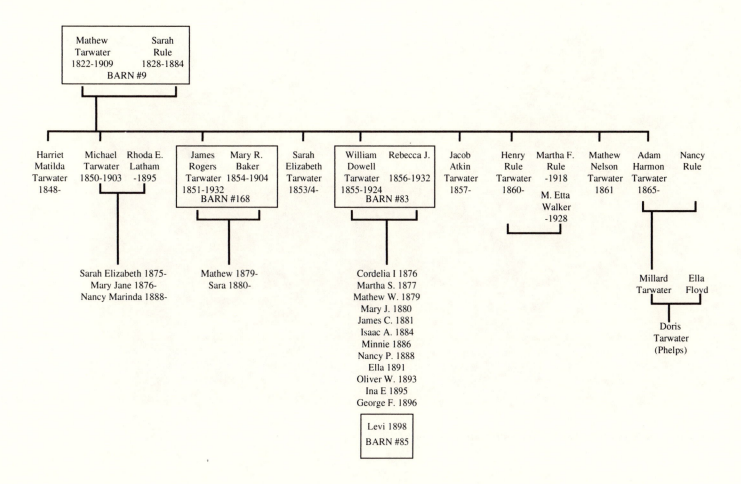

Figure 38. *Cantilever barn builders in the family of Matthew Tarwater.*

Figure 39. *Tarwater barn on Gists Creek, Sevier County (Sevier 9).*

Given the size of Matthew's family (his children were born between 1849 and 1866) and the fact that he lived to the considerable age of eighty-seven, farming with the assistance of his youngest son, a plausible case can be made to date construction in the interval from the late 1850s to the mid-1880s. Earlier dates would conform best with the general principle of barn dating. The disruption caused by the Civil War, while not so devastating in Sevier County as in other parts of Tennessee, may well have proven detrimental to barn raising. Matthew Tarwater's barn could well have been built in the 1870s when his sons were in their teens; or it is possible that his barn was indeed built in the 1880s, perhaps simultaneously with his sons' barns. It is impossible to say with the information available.

Matthew's second son, James Rogers Tarwater, married Mary M. Baker (1854-1904) and sired two children. His farm on Mathis Hollow Road included a cantilever barn (Sevier 168) built along almost identical lines to that of his father (figure 40). The probable date for James's barn is the 1880s, so the agricultural census data for 1880 are probably not representative of his farm's prosperity at a later period.

A comparison of James's farm with that of his brother, William Dowell Tarwater of Sevier 83, is instructive. From what can be determined in the agricultural census, the two brothers started from very similar conditions (see graph in Appendix A). Their barns, which are also very much alike, resemble their father's in general size. One interesting difference is that both brothers' barns are built with inner extensions of the primary beams to increase loft floor area, a feature absent in Matthew's barn. William and his wife, Rebeca J. (1856-1932), raised a very large family—fifteen children, of whom thirteen were living in 1900. The youngest of these, Levi, built a single-cantilever barn, Sevier 85, smaller than others on the Tarwater family farms.

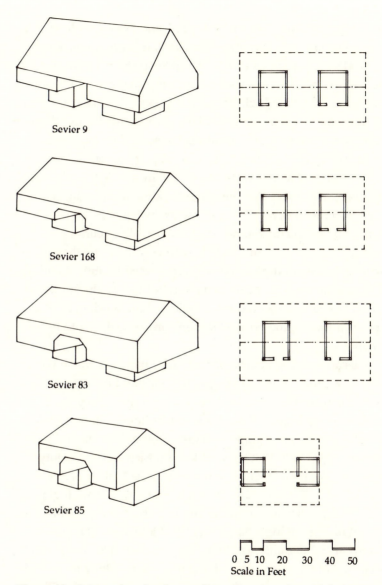

Sevier 9

Sevier 168

Sevier 83

Sevier 85

0 5 10   20   30   40   50
Scale in Feet

Figure 40. *Perspective sketches and comparative plans of the Tarwater barns along Gists Creek, Sevier County.*

The Ingle barns—Sevier 29, 89, 113, and 114—are physically close to one another, and they share similar proportions in detail (figure 41). All have inner extensions on the primary cantilevers. One feature that distinguishes the group is the secondary cantilever taper: on all the Ingle barns, the secondaries are shaped so that the beam's width is substantially greater that the depth (see figure 5). On Sevier 29, for example, the secondary cantilever measures 9 inches wide by 10 inches deep at the center, changing to a section 9 inches wide and only 5 inches deep at the end (figure 42). On Sevier 113, the corresponding dimensions are 10 inches by 12 inches, tapering to 9 inches by 5 inches. So pronounced a taper is uncommon and is particularly curious because the beam loses considerable stiffness by having these proportions. Local builders could not be expected to understand the structural basis for beam behavior, but one might think that practical experience would have indicated the wisdom of placing beams having the greatest dimension vertically, not horizontally.

We cannot assign particular builders' names to the Ingle barns. The progenitor, John Ingle (1812-87), and his wife, Celia (1813-79), moved before 1860 into Sevier County from North Carolina. Riley Ingle and Dr. Ingle, the builder's names given in the field, are the same person—a grandson of John named Riley Johnson Ingle (1875-1966), a well-known early Sevier County physician. We presume him to have been the builder of Sevier 29 and Sevier 88. Given that Dr. Ingle's father, William Ingle (1835-1917), was a farmer in the Tenth District, it is possible that he built one of the Ingle barns. In a 1990 interview, the ninety-three-year-old daughter of Riley Ingle recalled that "Dr. Ingle had a way of buying a place, fixing it up, then selling it. . . . The Ingle family lived in six different places in or near Sevierville during the time the children were growing up."[22]

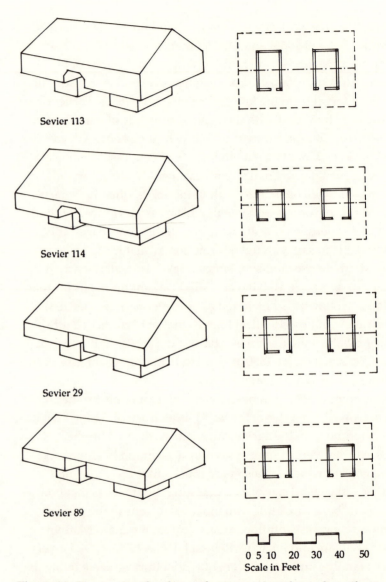

Sevier 113

Sevier 114

Sevier 29

Sevier 89

0 5 10   20   30   40   50
Scale in Feet

Figure 41. *Perspective sketches and comparative plans of cantilever barns built by the Ingle family, Ingle Hollow, Sevier County.*

Figure 42. *Ingle barn on Ingle Hollow Road, Sevier County (Sevier 29).*

Cantilever Barns among the Less Affluent

Most of the barns discussed thus far were built by farmers whose incomes were average to above average in the county. There were, however, less-prosperous farmers who also built cantilever barns. As one might expect, these families constructed smaller barns; single-cantilever types are found more often on poorer farms than on wealthier ones. James Bohanon (1836-1921), the builder of Sevier 60, represents one example of a less-prosperous farmer with a double-cantilever barn.[23] Although his farm along Upper Middle Creek Road was geographically close to the prosperous Middle Creek community, the land in his neighborhood consists not of gently rolling, well-watered acreage, but of steeper hillsides and a narrow valley

along the stream bed. Even today, the region along Upper Middle Creek presents difficult building sites, and as farmland it is definitely more challenging. Bohanon's grandfather, Henry Bohanon, was a Revolutionary War veteran from Virginia who held a land grant on the East Fork of the Little Pigeon River. His father, Henry Bohanon, Jr., married Catherine Powell, a Virginia native. James Bohanon is first found living along Upper Middle Creek Road in the 1870 census, when he and his wife, Clarinda Ogle, had five children. (The family would eventually include ten children.) James's father was born in Virginia, and Clarinda's mother was a native of North Carolina. In 1880, Bohanon was noted as being unable to read or write, and both his oldest child and his wife were unable to write.[24] The household in 1880 included seven children, ranging in age from two to seventeen; and James thought he was thirty-nine years old rather than the forty-four he actually was. His farm in 1870 consisted of thirty acres of improved land (expanded by 1880 to include two additional acres of apple orchards) and twenty acres of unimproved land. Farm value from 1870 to 1880 fell from $400 to $250, mirroring the drop in values countywide. His livestock value also declined, even though the number of animals remained much the same: two horses, two milk cows, seventeen sheep, and two pigs. His was also the smallest double-cantilever barn among those in Sevier County whose builders' names are known to us. The loft area of Bohanon's barn measures 829 square feet. By way of comparison, the Tarwater barns range in area from Matthew's 1,718 square feet to William D.'s 1,908 square feet; all are more than twice the size of Bohanon's barn.

John W. Whitehead (1842-78) is illustrative of the poorer single-cantilever barn builders.[25] The builder of Sevier 135, he farmed about fifty acres on LaFollette Road in the Third Civil District near several other surviving single-cantilever barns.

John's father, also named John, is listed in the 1850 agricultural census as Weishopt, doubtless the enumerator's attempt to write Weisskopf, German for "Whitehead." (This is a rare instance of a Germanic family name in Sevier County maintaining its original pronunciation.) By 1870, the name was given in its anglicized spelling, when the younger John is listed on his own as married with two children under two years of age. The census also notes that he was born in North Carolina and that his right to vote was abridged on grounds other than rebellion.

In 1878 John died, leaving his widow, Mary (1847-1924), to be listed in 1880 as managing the farm with nine children, aged one to eleven. She was unable to write, and the oldest child was illiterate despite attending school.[26] The oldest son, John C. Whitehead, turned twelve in June, yet his occupation was already listed as "farm laborer." Even when John the father was still alive, the family was not well-off. His personal property listed in the 1870 census was $250, at a time when his immediate neighbors were reporting four times that amount. One wonders if lack of money to pay a poll tax prevented him from voting. In 1880, the Whitehead farm consisted of thirty-seven acres of improved land and fifteen acres of unimproved land, the whole valued at $500, on which small amounts of wheat, corn, potatoes, apples, and peas were raised. Livestock included one milk cow, two other cattle, seven sheep, and ten pigs, for a total value of $50. By the standards of the times, this was a marginal farm, and not surprisingly the barn is correspondingly small. Its loft measured just a little over one thousand square feet. What is perhaps most remarkable is the survival to adulthood of all nine Whitehead children; by contrast in 1900, only one of the five children born to John C. Whitehead and his wife was still alive.[27]

Whitehead and Bohanon, two of the poorer farmers among owners of cantilever barns, share with Richard Reagan and many others of their generation the trait of illiteracy. Limited or no education meant that a considerable portion of the population had no access to books or farm journals that elsewhere might foster change. Just as new ideas would have difficulty getting into the region, local inventions like cantilever barns would have a small chance of getting out, at least in writing. In a very real sense, illiteracy contributed to the preservation of regional tradition, an aspect considered more fully in the following chapter.

CHAPTER 4

REGIONAL FACTORS

Even the cleverest invention will not be generally adopted if conditions for its acceptance are not right. In the case of cantilever barns, it appears that several qualities particular to East Tennessee promoted the development and perpetuation of this unusual building form. Self-sufficient agriculture, widely practiced in the counties under consideration, favored multipurpose barns. The forested regions supplied ample building material for hewn log and heavy timber frame construction at minimum cost. Physical isolation promoted by the region's geography worked against development of a market economy, thereby preserving older ways well into the twentieth century. Furthermore, cultural isolation reinforced conservative traditions, wherein locally developed ideas were repeated but not exported. All these factors acting in concert may explain why cantilever barns, once developed, flourished only within a limited area of East Tennessee.

Landscape and Agriculture

The settlers who moved into East Tennessee entered a land of rugged terrain that remains difficult to traverse even in the late twentieth century. While the region was still a wilderness, the mountains were an effective barrier to settlement; and even after European farmsteads were established, these hills served

to isolate the inhabitants from the larger world—a situation not entirely at odds with the Scots-Irish character. The counties of Tennessee in which cantilever barns have been found—Johnson, Carter, Washington, Unicoi, Greene, Cocke, Jefferson, Knox, Sevier, Blount, Meigs, and Bradley—are in the easternmost portions of the state, near the North Carolina state line, where the land is folded into ridges and valleys running roughly northeast to southwest along the line of the Appalachian mountains extending from New England to Georgia (see figure 3). Within the larger valleys of these counties are the major tributaries of the Tennessee River: the Holston, French Broad, Pigeon, Little Tennessee, and Nolichucky being the most important. Few cantilever barns are found in these fertile farming areas. Smaller streams that feed into these tributaries drain narrow valleys and coves, and it is on farms in this relatively rough and hilly country that the majority of cantilever barns were built. Unlike the broad river valleys, which had remained largely free of trees, this upland district was heavily forested, providing ample building material. Creating a homestead in the backwoods involved removing trees to create fields and pasture, and the wood thus obtained was employed for shelter, fuel, and the making of tools. Cantilever barns were sized to fit the scale of individual farms, and the barns' adaptability to multiple uses made them appropriate to self-sufficient farms.

Moving farther into the highlands, one finds the steepest land on the mountainsides was too sparsely settled and unproductive for agriculture to justify the size of a typical cantilever barn. Soil infertility combined with topography to make upland farms less productive for agriculture. The broader valleys, located on limestone formations, are generally characterized by loam and silt loam soil over red clay. "Valley lands of this type are considered to be of good natural fertility . . . [whereas] the soils of the rounded ridges . . . are recognized as much poorer than those of the valley lands."[1] The narrower ridge and valley lands are characterized by shallow shale soils, high in silt content, which are prone to clod and erode easily. They are also deficient in nutrients—nitrate, phosphoric acid, lime, and potash—making them only moderately productive under careful cultivation.

Rugged landscape and thin soil are accompanied in East Tennessee by a damp and humid climate. The upland areas receive an average of forty to eighty inches of rain annually, and particularly during the warm summer months the air is quite humid. Keeping livestock and stored crops dry is a continuing challenge, which the peculiar overhanging form of cantilever barns meets well. Animals in the cribs are better protected against driving rain than they would be in a log crib covered by a simple roof; wagons and other farm implements stored in the protected area around the crib's perimeter remain under cover while close at hand. The barn's form also works well for keeping dry the hay or other fodder stored in the loft. The central passage between the cribs functions in conjunction with the continuous vents below the eaves to promote air circulation through the upper floor—the passage serving as a breezeway,

and the ventilated open loft area drawing air upward like a chimney (figure 43). The gable roof, uncomplicated to build and maintain, sheds water effectively.

Furthermore, the cantilevered loft serves to keep the logs of the cribs dry, forestalling the rot which comes when wood is allowed alternately to be wet and dry. Provided that the barn is set on a slight rise so that ground water drains away from the building, the cribs of a double-cantilever barn will be surrounded by a band of dry earth from five to ten feet wide, created because rain drips off the roof at the same distance as the length of the cantilevers. This dry area becomes a deterrent to ground-dwelling termites, whose life-cycle requires continuous contact with moist soil and

Figure 43. *Loft interior showing continuous eave and gable end vents in the McMahan barn at Richardson Cove (Sevier 14).*

whose dietary staple is the cellulose of damp timber.[2] The cantilevered form thus effectively protects the log cribs from rot and termites, two major destroyers of wooden structures.

The architectural advantages of cantilever barns—relatively straightforward construction using readily available materials and commonly understood techniques to achieve superior protection for animals and fodder—would appeal to the self-sufficient farmers of East Tennessee, which may partially explain why cantilever barns were repeated, with variations, across the region. One variation, that of the double-height crib seen in several barns, may in fact have been a response to the introduction of tobacco culture in the late nineteenth century. The crop had been frequently grown in a small patch for home consumption, although not for export. In 1864, air-cured burley tobacco was first grown in Greene County, and during the 1880s its culture spread to the counties of upper East Tennessee. The transportation and market facilities necessary for large-scale tobacco farming did not come to Blount and Sevier counties until the 1920s, after the major period of cantilever barn construction was over, so its culture was probably not a factor in barn design there. (Needless to say, the lofts of cantilever barns serve admirably for drying the burley crop.) In Washington County, however, where burley tobacco culture began in the late nineteenth century, all four surviving cantilever barns have double-height log cribs that may represent the amalgamation of cantilever construction with traditional mountain tobacco-barn design. Simple gable-roofed log buildings were, from the very first, used as tobacco barns—both for air-dried and flue-cured leaves. Joseph Buckner Killebrew, at one time Tennessee's Secretary of the Bureau of Agriculture and an advocate of tobacco production in the state, described a typical flue-cured tobacco barn and its shortcomings:

There are many excellent farmers who contend that a log pen, twenty feet square, with four or five firing tiers, well-daubed, is the best of all houses for curing [tobacco. . . . However,] such barns rot down rapidly, and unless protected by bonnets or hoods on the ends, the injury done to the tobacco by beating rains will amount, in a few years, to the cost of a good barn.[3]

While not exactly a "bonnet or hood," the cantilevered upper level of these barns does effectively shelter the drying leaves from wind-driven rain. Spaces between the logs create a three-dimensional rack for inserting the poles on which the hands of tobacco are hung, and the use of these barns for tobacco would explain the absence of a loft floor (figure 44). Although dates of the Washington County barns are unknown, they may have been built in the 1880s when burley was becoming an important part of the economy. Positive evidence is lacking to link these or the two double-height log-crib cantilever barns in Sevier County specifically to tobacco cultivation.

Isolation

During the nineteenth century, when most cantilever barns were constructed, topography and cultural predilection combined to make the southern Appalachians a backwater rather than a frontier, inadvertently preserving folk elements, including cantilever barns, that otherwise might not have flourished. Self-sufficiency was the rule in all things, especially agriculture, and while some independent living was an extension of the highlander's character, at least a portion of it was imposed by geography. Except in the major river valleys, where watercourses were suitable for navigation and comparatively level ground could be found for roads, transportation was extremely

Figure 44. *Double-height log crib (Johnson 1).*

difficult. Roads were few, and trails that followed creek beds could become impassable in rainy weather. Had there been a cash crop or agricultural surplus, there would have been no economical way for it to be transported to market. Buckwalter's study of the pre-Civil War East Tennessee economy found that inadequate transportation was the major factor impeding the development of commercial agriculture in the region,[4] and the situation had not improved by 1874.

> It is a serious drawback to the farming interests of East Tennessee to have so few good roads. Usually they are execrable, and especially is this the case where the roads run transversely across the country. No successful efforts have been made to build turnpikes, though rocks are abundant and convenient for that purpose. With the exception of a few miles of McAdamized roads leading out of Knoxville, we believe there is not another in East Tennessee. . . . The tax the farmers indirectly pay in getting their produce to market over such roads is very burdensome, and the public mind should be directed to improvement in this particular. Wagons, passing over such roads as prevail in East Tennessee, soon wear out and break down, and teams are strained and overtaxed without doing more than half the work that they might do on smooth roads.[5]

Numerous travel accounts from the period before and after the Civil War attest to the primitive state of mountain roads.[6] Although in 1858 the East Tennessee and Virginia Railway completed its route through the Tennessee Valley, providing a rail connection from Bristol, Virginia, to Alabama, Georgia, and

South Carolina, getting from the backwoods communities to the railroad was still difficult. Spur tracks laid for specific industries, such as coal mining in the Cumberlands or timbering in the Unaka range, improved transportation options, but only two railroads eventually crossed the mountains into North Carolina, one route following the French Broad from Morristown through Newport to Asheville, and the other following the Nolichucky through Erwin. Neither route passed through Blount or Sevier counties, the heart of cantilever barn territory.

For those sections of East Tennessee not served by rail, river transportation was to be the most expedient mode for moving people and bulk cargo. During the 1890s, Sevier County's newspaper, *The Republican Star*, contained frequent mention of the height of the Little Pigeon River: too much water made navigation dangerous, while low water rendered it impossible. Under normal conditions, flat boats on the Little Pigeon River transported passengers and cargo from Sevierville to Catlettsburg, where steamboats on the French Broad made the trip to either Knoxville or Dandridge. Ten hours' travel was required to reach Knoxville (a distance of about twenty-five miles), and about six hours were needed to get to Dandridge; but even so, the water route was easier and faster than going overland.

> In 1900 it took one and a half days to travel from Sevierville to Knoxville by wagon. . . . The Sevierville Pike, which ran twenty-eight miles from Sevierville to Knoxville by way of Boyd's Creek, was a "modern" macadam thoroughfare, recently completed in 1900. . . . Other roads had no gravel and, depending on the amount of traffic carried, some were little better than dirt paths. With the great

amount of rainfall the county receives, these roads could turn quickly to mud and in the summer were dry and dusty. There were few bridges across the county's many rivers and streams. Wagons, buggies, and those on horseback had to ford these obstacles, while footlogs were available to walkers.[7]

Understanding the transportation obstacles faced by East Tennessee residents is important because it helps to explain the isolation experienced in the back country. This isolation contributed in at least three ways to the culture of cantilever barns. First, isolation worked against the development of a market-based farm economy oriented toward cash crops. Because there was such poor access to markets, self-sufficiency remained the dominant farm mode and cantilever barns provided or continued to provide adequate accommodation for farm needs. Second, isolation compounded by illiteracy insulated farmers from "progressive" practices which recommended the construction of single-purpose farm structures and expressed scorn for log construction.[8] In the absence of information about their barns' inefficiencies and perceived backwardness, East Tennessee farmers continued to build and use cantilever barns. The Agricultural Extension Service's encouragement from the 1920s onward to build specialized barns for specific animals and crops instead of multipurpose barns went largely unheeded in East Tennessee.

Last, isolation contributed to an inwardly focused, conservative community, which was more likely to use or adopt ideas developed locally than to look abroad for new ideas. While the rest of the nation was experiencing prosperous expansion, Appalachian families practiced the living patterns established by the early settlers. Appalachian religious and cultural differences became barriers which inhibited the highlanders' absorption by the outside world, but which also preserved speech, folk songs, and other ways of life long since forgotten in the nation as a whole. After about 1850, migration into the region virtually ceased, despite promotion of the state as a haven for European peasantry. Some East Tennesseans emigrated to more fertile lands in western Tennessee, Arkansas, Missouri, and Texas. In rural districts, population levels thus grew slowly during the nineteenth century and then stayed nearly constant or grew only slightly from 1900 to 1950. As a cultural eddy in the larger nineteenth-century world, East Tennessee maintained until about 1920 a way of life little changed since the Civil War. Cantilever barns were part of that preserved culture, and they too survived.

CHAPTER 5

THE BARNS OF MIDDLE CREEK

The Middle Creek community, located in the center of Sevier County, has a substantial and varied collection of cantilever barns. Within a five-mile stretch along two county roads, nineteen cantilever barns have survived, giving this community one of the highest concentrations of barns found anywhere in our survey. The Middle Creek barns comprise all cantilever types found in the larger survey except for the four-crib type. By taking Middle Creek as a case study, it is possible to demonstrate how the factors affecting cantilever barns—ethnic background, agriculture, topography, social and cultural context, and family connections—had a collective impact on the barns in one specific area. Our investigation of the barns of Middle Creek was assisted by the chance survival of some papers related to Isaac Trotter, one of the barn builders; the knowledgeable recall of the late Glenn McMahan, descendant of another barn builder; and most particularly the historical expertise of the county historian, Beulah Duggan Linn, whose ancestral roots are connected to Middle Creek. These, in conjunction with population census records, newspaper items, and agricultural census data, have allowed reconstruction of the probable settlement sequence and farm economy out of which cantilever barns were built in Middle Creek.[1]

Historical and Agricultural Profile of Middle Creek

Middle Creek was one of the earliest settlements in the county, perhaps because its gently rolling lands were conducive to farming and the creek provided a ready water supply. The Middle Creek community centers on two main through roads: the north-south Middle Creek Road, which in the nineteenth century connected Sevierville, the county seat, and the settlements of Pigeon Forge and Stinnett; and Roberts Schoolhouse Road (also known as Jayell Road), which extends roughly east-west from Middle Creek Road to the Pittman Center Road running parallel to the Little Pigeon River (map 5).

Although the first official land grants in the district were made in 1808, European settlement in the area was already a quarter-century old by this time. Local tradition recounts the presence of a fort or blockhouse begun by the family of Robert Shields in 1784-85 on a site near the present Middle Creek Methodist Church. According to later accounts,

Shield's Fort when completed was a long building, 16 x 100 feet with low ceiling and attic. It was constructed of heavy logs with a fireplace at each end. There were four outside doors, several window openings without glass, and numerous portholes at convenient places, upstairs and down. The original building contained living quarters for six families, with a large common kitchen at one end, a common living room at the other. The building was in the midst of an oblong yard of about a quarter-acre surrounded by walls 12 feet high. The walls consisted of double rows of logs standing on end, closely spaced, sharpened at the top, and fastened together with wooden pins—the spring was within the enclosure, as were stables for stock, and all other buildings. Fortunately for the Shields Family the fort was a dozen miles off the regular Indian trails and it was never attacked by large war parties. It was frequently disturbed by small roving bands of two or three Indians, who might fire from ambush at those working in the clearing.

Nearly four years were required to complete the original structure. It was in the Fort that Robert Shields and his children and grandchildren lived for nearly 20 years. Seven of his sons brought wives to the Fort. As the family circle grew, the size of the stockade was increased. Eventually the McMahan and some of the Shields boys moved into their own separate quarters nearby. Both Robert Shields and his wife, Nancy Stockton, died at this Fort about 1805, exact date unknown, and were buried in unmarked graves on a nearby hillside.[2]

The site of the fort was later the location of religious campground meetings held from the 1830s. In addition to Middle Creek Methodist, two other churches served the community, Roberts Memorial Methodist (built 1857 as a schoolhouse and

Table 2. *Agricultural Census Data for the Fourth Civil District of Sevier County in 1880.*

Classification of Farms according to Tenure

Fourth District		Tenure	Sevier County	
Number	Percentage		Number	Percentage
86	67	Cultivated by owner	1530	76
20	15	Rented for fixed amount	51	3
26	18	Rented for share of the crop	433	21

Classification of Fourth District Farmland according to Ownership

Fourth District		Tenure
Acreage	Percentage	
6283	80	Cultivated by owner
401	5	Rented for fixed amount
1086	15	Rented for share of the crop

Average Farm Size and Value according to Tenure

	Fourth Civil District Farms			Sevier County
	Cultivated by owner	Rented for fixed amount	Rented for share of crop	
Improved land (acres)	73.0	20.0	46.3	42.6
Unimproved land (acres)	93.6	45.7	110.6	100.6
Farm size (acres)	166.6	65.7	154.9	143.25
Farm value	$ 1700.00	$ 470.00	$ 1072.00	$ 950.00
Value of farm machinery	$ 77.00	$ 20.00	$ 35.00	$ 42.00
Value of livestock	$ 263.00	$ 149.00	$ 147.00	$ 176.00

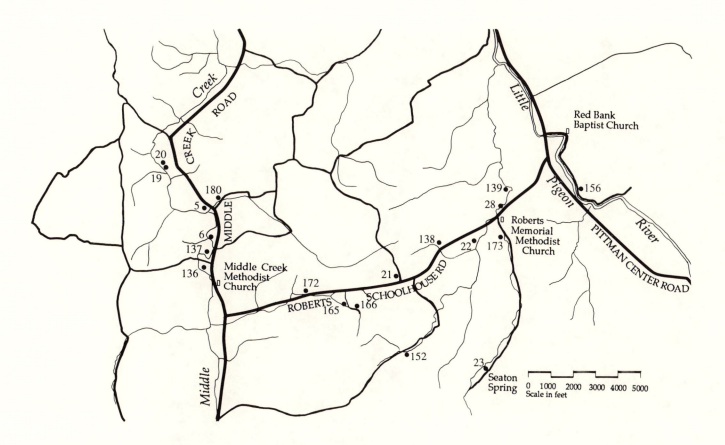

Map 5. *Middle Creek community of Sevier County, with dots indicating the locations of surviving cantilever barns.*

church), and Red Bank Baptist Church (founded 1840).[3] The churches and their associated grammar schools (which met in the church buildings) were central to the social life and definition of the community. Politically, the area comprises part of the Fourth Civil District of Sevier County.[4]

By the late nineteenth century, farms here were more pros-
perous than the averages for the county (table 2). Statistics based on the 1880 agricultural census show that while the size of owner-occupied farms in the Fourth Civil District is not significantly above the county average, the acreage under cultivation on these farms is 1.7 times the county average and the value of owner-occupied farms is 1.8 times larger.

Figure 45. *Aerial photograph of Middle Creek, flown in 1936.*

To a large extent, the Middle Creek area remains agricultural even today, and the land-use pattern does not appear to have undergone significant shifts within the last half-century. A comparison of wooded and open lands as shown on the earliest aerial photographs (figure 45) flown in 1936 with the most re-cent United States Geological Survey topographic maps of the region (map 6) reveals that the location of wooded and open plots has changed only slightly in the last sixty years. The major difference has been more houses located along the roads as sub-divisions have begun to replace fields.

Map 6. *Middle Creek as seen in 1972 United States Geological Survey map.*

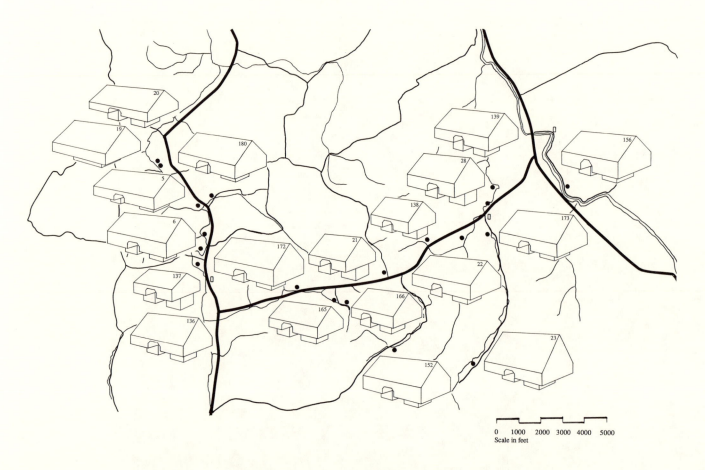

Map 7. *Middle Creek community, showing the nineteen surviving cantilever barns.*

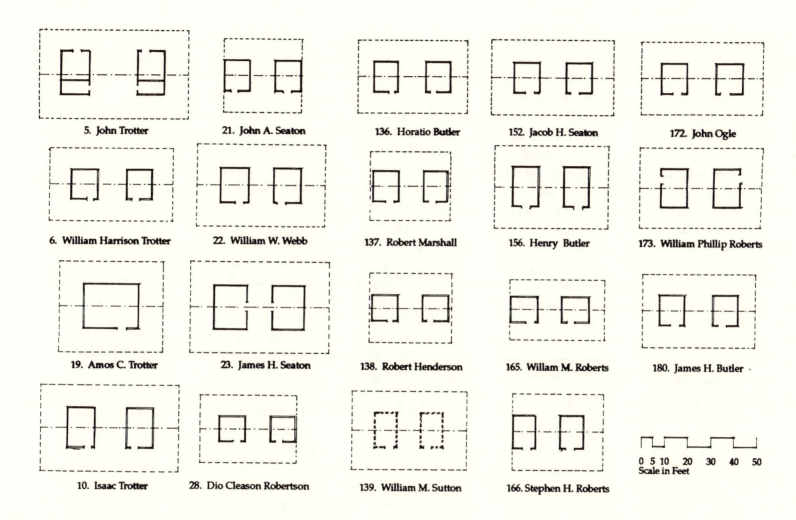

Figure 46. *Comparative plans of the Middle Creek barns.*

The Trotter Family Barns along Middle Creek Road

Family interrelationships are important to an understanding of the probable connections among the barns that were built. All seven cantilever barns along Middle Creek Road are associated with the Trotter family, and as these appear to be the first ones built in the community, we will be consider them first (figure 47). By 1811, the Maryland-born progenitor of the family, John Trotter (1777-1856), held a land grant for one hundred acres at the northern end of Middle Creek Road. The boundaries of his grant, like those of his neighbors, are decidedly irregular, owing to the fact that land was claimed on the ground first and surveyed later; Trotter's tract is more understandable when considered in light of its topography than when represented by a surveyor's plat. Adjoining land grants by the waters of Middle Creek were issued to Charles Clabough, Sr., James Mathis, Meddy White, Nathan Veach, and Jennet Tipton, but the Trotter family eventually acquired most of their holdings (map 8).[5]

Early in the nineteenth century, John Trotter married Meddy White's daughter, Asa White (1782-1854), who had been born in Bertie County in northeastern coastal North Carolina. The large two-crib double-cantilever barn designated as Sevier 5 is located on property originally belonging to Meddy White (1760-1816/18); White's log house stood across the road. Popular accounts hold that the barn was built by White, but more likely it was constructed by John Trotter about 1820 and is thus the oldest cantilever barn in the area.

This barn presents a puzzle. On one primary cantilever, "BUILT 1781" is inscribed in an antique hand. Could this be an authentic date? Much as one would like to believe it to be true, we are inclined to doubt this unusually early attribution. Common sense, rather than documentary evidence, suggests a later date. How could White have built this barn in 1781 when his daughter was born the following year over 400 miles away? This barn is quite large, appropriate to the farming activity of John Trotter. Were it to have been built in 1781, during the period when all European settlement in the region was unauthorized, it would have reflected a more considerable farming enterprise than we believe existed in the region at that time. One possible explanation for the inscribed date is that the barn reused some of the logs from Shields's fort, which was dismantled when its usefulness had passed.[6] The Trotter barn has unusually large beams, which do not taper at the ends of the cantilevers, suggesting that the timbers may have been cut for some other purpose and then recycled. The 1781 date may thus refer to the original fort construction, allowing for some historical backdating. (It is also possible that the barn was built by Trotter's son, William Harrison Trotter [1814-87], a prosperous man who built the adjoining house in 1848.) Meddy White himself remains a shadowy figure in county history. Aside from being expelled in 1811 from fellowship in the church for consumption of alcoholic beverages, he leaves little trace of his passing.

The Trotter barn's generous proportions and sizable timbers bespeak a substantial farm and a confident builder. None of the other Middle Creek barns (and few of all the barns surveyed) equal its overall size or its number of secondary cantilevers (eighteen). Its overall proportions of width to depth are reflected in the proportions of four other Middle Creek barns, Sevier 6, 137, 152, and 172 (compare the plans on figure 46). Unlike all other cantilever barns in the Middle Creek area, however, the Trotter barn's primary cantilevers extend five feet into the central aisle, reducing the wide aisle to an effective width of ten feet five inches. Each of the cribs is subdivided transversely into two pens of unequal size, a feature seldom found in cantilever barns and nowhere repeated in Middle Creek. Another

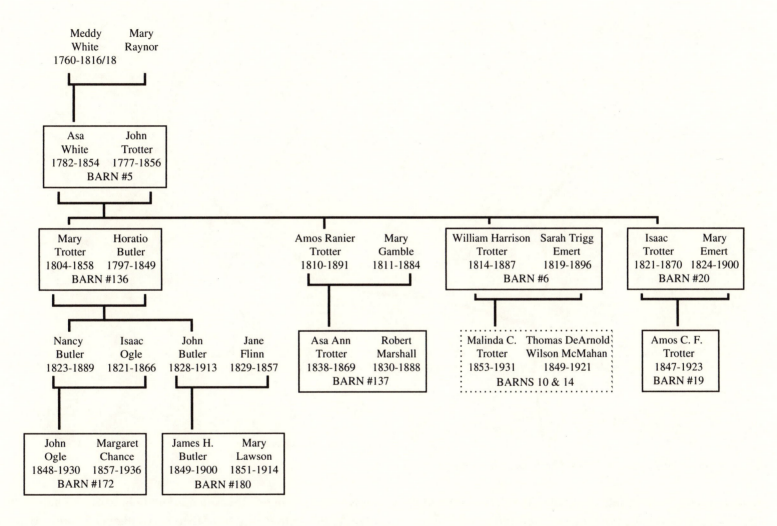

Figure 47. *Family connections among builders of cantilever barns living along Middle Creek Road. Sevier 10 and 14 are located at Richardson Cove, but these barns are connected by family ties to Middle Creek.*

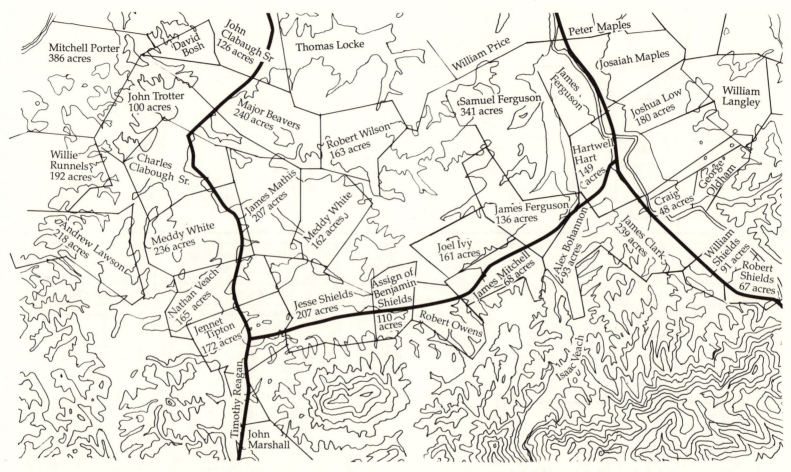

Map 8. *Original land grants in the Middle Creek community.*

anomaly occurs in the relationship of crib to loft. Although the cribs are the same size, the cantilever projections to the four sides are not, so that the loft floor is not centrally placed over the cribs (figure 48). The barn originally featured a threshing floor, but the continuous sill logs on which this rested have

now been removed. This splendid early barn is also linked by family ties to the large McMahan barns (Sevier 10 and 14) at Richardson Cove (see figure 12). John Trotter's granddaughter Malinda Trotter married T. D. W. McMahan, the builder of the Richardson Cove barns.

Figure 48. *Log cribs and cantilevers of the Trotter barn on Middle Creek Road (Sevier 5). An extensive addition masks the barn's original cantilevered form.*

According to the 1850 census of agriculture, John Trotter farmed seventy-five acres of land and held another seventy-five acres of woodland; the cash value of the farm was set at $600. He owned two horses, three cows, four sheep, and eight swine; his crops included wheat, corn, buckwheat, oats, rice, flax, hay, peas, Irish potatoes, and sweet potatoes. His farm manufactures, including two hundred pounds of butter, sixty pounds of beeswax, and twelve pounds of cheese, were probably used primarily for barter at the local store (table 3).[7]

John Trotter's land holdings along Middle Creek became

the foundation of his children's farms. In 1822, his eldest daughter, Mary (1804-58), married Horatio Butler (1797-1849), whose father also had come from Maryland. At the southern end of Trotter's land, Horatio Butler built a two-crib double-cantilever barn (Sevier 136) in about 1830 on what was once the property of Nathan Veach. Butler's two-story log house still stands adjacent to the barn, facing Center View Road.[8] His holdings at the time of his death were impressive: three hundred acres of improved farmland and eight hundred acres of woodland; eight horses, five cows, ten other cattle, twenty sheep, and thirty pigs. The crops Butler raised were essentially the same as those found on Trotter's farm, except that he did not grow rice but did raise forty pounds of tobacco. Butler's barn is smaller than Trotter's, having only fourteen secondary cantilevers, but its proportions of length to width are very similar. It lacks the inward extension of the primary cantilever found on Trotter's barn, and the inner faces of the primary cantilevers are notched to secure the skids which would be used to manipulate the secondary cantilevers into position.

In 1854, John Trotter deeded his farm to his eldest son, Amos Ranier Trotter (1810-91), in exchange for lifetime care of himself and his blind daughter. After John Trotter's death in 1856, Amos moved to neighboring Blount County, where his wife's family lived, and the Trotter holdings along Middle Creek became the property of John's two younger sons, William Harrison Trotter (1814-87) and Isaac Trotter (1821-70). Each of these sons built a cantilever barn. William H. Trotter held land immediately to the south of his father's homestead, and in about 1845 he built a two-crib double-cantilever barn (Sevier 6) within sight of his father's barn.[9] It is larger than Butler's barn, having the same number of secondary cantilevers, but not quite as large

Table 3 column headers: Census Year | Improved acreage | Unimproved acreage | Cash value of farm | Value of machinery | Horses | Asses and mules | Milch cows | Working oxen | Other cattle | Sheep | Swine | Value of livestock | Wheat (bushels) | Corn (bushels) | Oats (bushels) | Wool (pounds) | Irish potatoes (bushels) | Sweet potatoes (bushels) | Butter (pounds) | Hay (tons) | Molasses (gallons) | Honey (pounds) | Wood cut (cords) | Other

Census Year	Imp. acre.	Unimp. acre.	Cash value	Value mach.	Horses	Asses/mules	Milch cows	Working oxen	Other cattle	Sheep	Swine	Value livestock	Wheat	Corn	Oats	Wool	Irish pot.	Sweet pot.	Butter	Hay	Molasses	Honey	Wood	Other
John Trotter (1777-1856) Barn 5																								
1850	70	75	600	150	2		3		2	4	8	209	40	550	30	7	12	30	200			60		75 lb rice; 12 lb cheese.
Horatio Butler (1797-1849) / Mary T. Butler (1804-1858) Barn 136																								
1850	300	800	1500	150	8		5		10	20	30	450	75	800	100	30	25	75	100					20 lb flax; 40 lb tobacco.
William Harrison Trotter (1814-87) Barn 6																								
1850	94	473	700	72	3	3	3	2	3	7	17	480	40	300	75	25		75	50	1				13 bu buckwheat; 20 bu peas.
1860	200	330	6000	500	6		3	2	4	21	100	1385	300	1200		50	5	10	50		5			5 bu peas.
1870	175	250	6000	500	6	2	4		15	19	35	930	250	400		25		35	150	10	40	50		
1880	123		5000	175	3	3	4		8	13	17	553	375	1000		50			365	2			50	8 acres orchards.
Isaac Trotter (1821-70) / Amos C. F. Trotter (1847-1923] Barns 19 and 20																								
1850		12	25	75	2	1	2		11	5	8	200	50	400	150		30	30	30					17 bu buckwheat.
1860	755	110	2800	200	3		4	2	5	28	35	675	120	600	60	60			25		120			35 bu rye.
1870	70	130	4500	400	4	1	4		12	33	12	1140	248	600		40	6	25	200	8	30	100		
1880	100	96	3350	253	3	6	3		12	12	10	510	369	700					365	11		165		
Robert Marshall (1830-88) Barn 137																								
1860	50	70	800	75	3		2		2		25	330	40	500		40		100	100	3	100			
1870	100	100	800	15	1		3		2	17	12	100	65	200		20			100	2	65			
1880	80	125	1000	25	2		1		5		10	200	160	300					25	4			50	25 bu apples.
John Ogle (1848-1930) Barn 172																								
1880	38	77	600	10	1	1	1		1		8	165	86	300					30				25	1 acre apple orchards.
Averages for Owner-occupied Farms in the Fourth Civil District																								
1880	73	94	1700	77								263												

Table 3. *Agricultural Census Statistics for Barn Builders along Middle Creek Road.*

as the first Trotter barn. Isaac Trotter's farm was on the northern end of John Trotter's land, and his two-crib double-cantilever barn (Sevier 20) was probably built about 1855.[10] The 1850 agricultural census indicates that Isaac was farming land jointly with William H. Trotter, but over the next decade his land holdings grew from 12 to 865 acres, probably due to inheritance from his father's estate. He eventually built a cantilever barn almost as large as that of John Trotter.

We know a fair amount about the activities and occupations of William H. and Isaac Trotter. They married women who were first cousins to each other. Both men were politically active in the county, William serving as county trustee from 1842 to 1852,[11] and Isaac as a justice of the peace[12] and chairman of the Sevier County court in 1862, using this position to claim exemption from military service during the Civil War. Through shrewd purchases of land, William H. Trotter became wealthy, and in 1848 he built a sizable house that still stands near the John Trotter barn.[13] William H. Trotter also served as a physician, although it is uncertain whether he received formal training in medicine.[14]

The house associated with the Isaac Trotter barn is said to have been constructed in 1844, a date that seems a decade too early according to the agricultural census records for Isaac. The Isaac Trotter farm contains another barn in addition to the two-crib double-cantilever Sevier 20. This second barn, Sevier 19, is also doubly cantilevered, but it features only one crib, a two-story log structure (figure 49). This barn is unlike any other cantilever barn in Middle Creek, not only for its overall form but also for its details, particularly its use of V-notching. The general design of two-story log cribs links it to similar two-story log cribs on double-cantilever barns in Greene and Johnson counties, suggesting the form of tobacco barns; and it is possible that this barn was erected by Isaac Trotter's son and heir, Amos

C. F. Trotter (1847-1923), in about 1895 for hanging tobacco. Agricultural census data is unfortunately not available to confirm the hypothesis that A. C. F. Trotter grew substantial amounts of burley tobacco, but it is known that cultivation of tobacco in East Tennessee increased about the turn of the century, and it would be consistent with Trotter family farming practices to be among the market-oriented farmers growing burley tobacco.

Two other cantilever barns remain along Middle Creek Road, both with connections to the Trotter family. Robert Marshall (1830-88) is credited with building Sevier 137 about 1860.[15] Marshall married Asa Ann Trotter, the daughter of John Trotter's oldest son, Amos Ranier Trotter; and his farm was located between the holdings of William H. Trotter and Horatio Butler. This barn is one of three single-cantilever barns in Middle Creek. Its cribs are close to the same size as those of its neighbor, Sevier 6, although the overall size of the barn is considerably smaller than a double-cantilever barn. While technically it is a single-cantilever barn, in fact the primary cantilevers project eight inches from the sides of the supporting cribs. An unusual braced purlin structure is used for the loft frame. Agricultural census records show Marshall's farm to be smaller than those of his neighbors. He owned about 200 acres of land, of which less than half was under cultivation. By comparison, W. H. Trotter owned between 425 and 530 acres, with 175 to 200 acres under cultivation. Like his neighbors, Marshall grew corn, wheat, sorghum, sweet potatoes, and apples; his livestock included horses, cows, pigs, sheep, and poultry.

The seventh barn on Middle Creek Road, Sevier 180, was built after 1880 by James H. Butler (1849-1900), grandson of Horatio Butler. Not surprisingly, James H. Butler's barn matches the Horatio Butler barn very closely in its overall

Figure 49.
Double-height log crib of the Trotter barn (Sevier 19).

dimensions in both plan and elevation. The chief difference be-tween the two barns comes in the number of secondary cantilevers, the later barn having only ten instead of fourteen. James H. Butler does not appear in the 1880 census of agriculture, so no information about his farm is known.

Cantilever Barns along Roberts Schoolhouse Road

Along Roberts Schoolhouse Road, there are twelve cantilever barns, ten of which are connected through marriage to three families, Roberts, Robertson, and Seaton, with the remaining two having ties to the Trotter family (figure 50). Of the twelve,

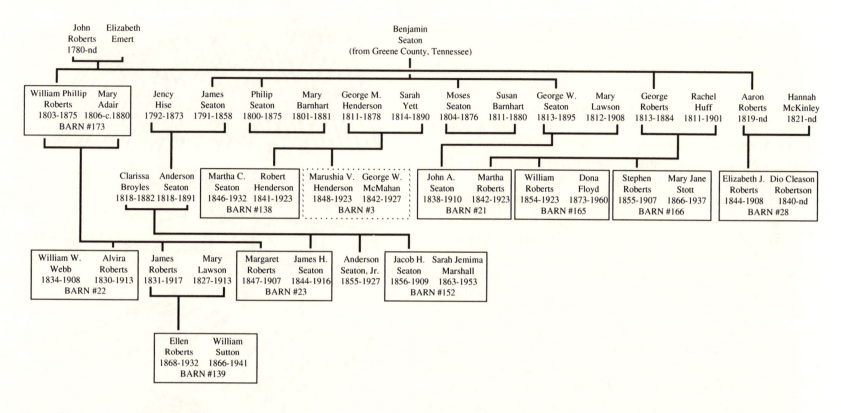

Figure 50. *Family connections among cantilever barn builders living along Roberts Schoolhouse Road. Sevier 3 is located north of Sevierville, but it is connected by family ties to the Middle Creek area.*

the oldest appears to be Sevier 173, constructed about 1860 by William Phillip Roberts (1803-75), the same man who donated land in 1857 for construction of the log schoolhouse (and Methodist Church) which bears his name.[16] Roberts's barn is among the largest in the district, and, like the John Trotter barn, has some anomalies that are not repeated in neighboring barns. The most notable peculiarity is the placement of crib openings: alone of all the barns surveyed, these crib doors open to the sides of the barn, which means that the initial log course was laid along the sides of the crib, not along the front and back as is generally the case. Corner timbering for the cribs is a mixture of V - and half-dovetail notching, with the V-notches used for the first log course only. Studies by Jordan and others have found that this is a not uncommon feature of log cabins, but its occurrence in cantilever barns is relatively rare. The crib timbers are hefty, the largest measuring eighteen inches high by nine inches wide.

Roberts had a large farm. The 1860 agricultural census shows that he owned 450 acres, of which 160 were under cultivation. Much of his wealth was in livestock, including nine horses. Of the Middle Creek cantilever barn owners, only William H. Trotter had more valuable livestock (table 4).

The William Phillip Roberts barn is located just off Roberts Schoolhouse Road on Seaton Springs Road. Traveling a mile and a half down Seaton Springs Road brings one to the end of the valley and Sevier 23, which most closely relates in form to the Roberts barn (figure 51). Sevier 23 was constructed about 1875 by Reverend James H. Seaton (1844-1916), a Methodist minister and proprietor of Seaton's Summer City, a local resort organized around Seaton Spring.[17] James Seaton married a daughter of William Phillip Roberts, providing both a familial and geographic link to Sevier 173. To accommodate guests, Seaton built a two-story house or inn at the head of the valley, and the barn is situated just behind. The barn's size in plan makes it the largest in the Roberts Schoolhouse district, and the unusual height of its loft gives it the greatest volume of any barn in the neighborhood (figure 52). Two other atypical features should be noted: its cribs open to the center aisle, not the front, and square notches are used for corner timbering on the cribs. Substantial timbers were available for the barn's construction: the largest crib log measures twenty inches high by seven inches wide. From agricultural census data, James H. Seaton's farm does not appear substantial enough to justify such an enormous barn.

Figure 51. *Exploded perspective explaining the construction of the Seaton barn at Seaton Springs (Sevier 23).*

Table 4 — Agricultural Census Statistics for Cantilever Barn Builders along Roberts Schoolhouse Road.

William Phillip Roberts (1803-75) Barn 173

Census Year	Improved acreage	Unimproved acreage	Cash value of farm	Value of machinery	Horses	Asses and mules	Milch cows	Working oxen	Other cattle	Sheep	Swine	Value of livestock	Wheat (bushels)	Corn (bushels)	Oats (bushels)	Wool (pounds)	Irish potatoes (bushels)	Sweet potatoes (bushels)	Butter (pounds)	Hay (tons)	Molasses (gallons)	Honey (pounds)	Wood cut (cords)	Other
1850	78	400	800		7		4	2	8	42	25	355	40	400	200	6	15	10	50					
1860	160	290	2000	30	9		2		2	21	22	1155	67	600		40		60	75			55		
1870	165	265	3000		5		2		4	20	7	682	145	500	40	50			300	3	30			

Henry Butler (1826-1916) Barn 156

Census Year	Improved acreage	Unimproved acreage	Cash value of farm	Value of machinery	Horses	Asses and mules	Milch cows	Working oxen	Other cattle	Sheep	Swine	Value of livestock	Wheat (bushels)	Corn (bushels)	Oats (bushels)	Wool (pounds)	Irish potatoes (bushels)	Sweet potatoes (bushels)	Butter (pounds)	Hay (tons)	Molasses (gallons)	Honey (pounds)	Wood cut (cords)	Other
1850				10	2	2	2		4	12	15	230	38	400	75	15	20	20	100					2 bu buckwheat.
1860	100	60	3500	200	5		2	1		8	75	890	200	1000	75	16								125 lb tobacco.
1870	200	260	7000	150	6		4		6	30	28	690	330	600	50	30	40	30	210	15	32			Paid $200 wages.
1880	152	300	6000	400	3	8	4		20	15	15	1000	535	800	600	19			500	15		40	20	7 bu seed; 100 lb cheese.

William W. Webb (1834-1908) Barn 22

Census Year	Improved acreage	Unimproved acreage	Cash value of farm	Value of machinery	Horses	Asses and mules	Milch cows	Working oxen	Other cattle	Sheep	Swine	Value of livestock	Wheat (bushels)	Corn (bushels)	Oats (bushels)	Wool (pounds)	Irish potatoes (bushels)	Sweet potatoes (bushels)	Butter (pounds)	Hay (tons)	Molasses (gallons)	Honey (pounds)	Wood cut (cords)	Other
1860	3	100	300	5	2		1		2	5	15	105	52	200	50	5	10	10	50		30			11 bu peas.
1870	17	32	570	100	2		4		6	10	11	400	40	150	35	15		12	50	1	10			
1880	125	235	1200	100	5		2		7	20	50	500	235	600	20	40			100			9	25	

John A. Seaton (1838-1910) Barn 21

Census Year	Improved acreage	Unimproved acreage	Cash value of farm	Value of machinery	Horses	Asses and mules	Milch cows	Working oxen	Other cattle	Sheep	Swine	Value of livestock	Wheat (bushels)	Corn (bushels)	Oats (bushels)	Wool (pounds)	Irish potatoes (bushels)	Sweet potatoes (bushels)	Butter (pounds)	Hay (tons)	Molasses (gallons)	Honey (pounds)	Wood cut (cords)	Other
1880	25	21	400	12	2		1		1	2	9	175	41	150	10	4			20			40		

Dio Cleason Robertson (1840-nd) Barn 28

Census Year	Improved acreage	Unimproved acreage	Cash value of farm	Value of machinery	Horses	Asses and mules	Milch cows	Working oxen	Other cattle	Sheep	Swine	Value of livestock	Wheat (bushels)	Corn (bushels)	Oats (bushels)	Wool (pounds)	Irish potatoes (bushels)	Sweet potatoes (bushels)	Butter (pounds)	Hay (tons)	Molasses (gallons)	Honey (pounds)	Wood cut (cords)	Other
1880	87	25	2500	75	5	1	3		5	13	30	470	200	500	100				150				10	30 bu apples.

Robert Henderson (1841-1923) Barn 138

Census Year	Improved acreage	Unimproved acreage	Cash value of farm	Value of machinery	Horses	Asses and mules	Milch cows	Working oxen	Other cattle	Sheep	Swine	Value of livestock	Wheat (bushels)	Corn (bushels)	Oats (bushels)	Wool (pounds)	Irish potatoes (bushels)	Sweet potatoes (bushels)	Butter (pounds)	Hay (tons)	Molasses (gallons)	Honey (pounds)	Wood cut (cords)	Other
1870	47	160	2000	118	2	1	2		2	20	13	343	187	400						1	14			
1880	76	287	2200	90	5		2		7	14	37	351	150	600		18	9	20	150		33	50	50	

James H. Seaton (1844-1916) Barn 23

Census Year	Improved acreage	Unimproved acreage	Cash value of farm	Value of machinery	Horses	Asses and mules	Milch cows	Working oxen	Other cattle	Sheep	Swine	Value of livestock	Wheat (bushels)	Corn (bushels)	Oats (bushels)	Wool (pounds)	Irish potatoes (bushels)	Sweet potatoes (bushels)	Butter (pounds)	Hay (tons)	Molasses (gallons)	Honey (pounds)	Wood cut (cords)	Other
1870	50	50	800	40	2		2	2	2	16	11	310	30	75	30	25					10	24		
1880	70	10	200	8	2	1						192	57	42										30 bu apples.

Jacob H. Seaton (1856-1909) Barn 152

Census Year	Improved acreage	Unimproved acreage	Cash value of farm	Value of machinery	Horses	Asses and mules	Milch cows	Working oxen	Other cattle	Sheep	Swine	Value of livestock	Wheat (bushels)	Corn (bushels)	Oats (bushels)	Wool (pounds)	Irish potatoes (bushels)	Sweet potatoes (bushels)	Butter (pounds)	Hay (tons)	Molasses (gallons)	Honey (pounds)	Wood cut (cords)	Other
1880	67	440	1500	85	2		3		5	23	11	300	50	500		23			300	5		25	20	225 bu apples & peaches.

Table 4. *Agricultural Census Statistics for Cantilever Barn Builders along Roberts Schoolhouse Road.*

Figure 52.
*Seaton barn
at Seaton
Springs
(Sevier 23).*

One may surmise that he stocked it with material purchased from others, including his father, Anderson Seaton, to supply the needs of his guests and their animals. Together, the house and barn still present a handsome image of a well-built and soundly maintained farmstead.

Back on the main road is another Seaton barn, Sevier 21,

built after 1870 by James's cousin, John A. Seaton (1838-1910).[18] It is a single-cantilever barn, as is its neighbor and contemporary, Sevier 138 of Robert Henderson (1841-1923).[19] Both follow rather closely the dimensions and proportions of the Marshall barn (Sevier 137) on Middle Creek Road, the Henderson barn also having a one-foot projection of its primary cantilevers to the outer side of the cribs. A log cabin still exists on the Seaton property.

Sevier 138 has square cribs, a feature it shares with its neighbor to the east, Sevier 28 of Dio Cleason Robertson (1840-ca. 1900), which is one of the three half-double-cantilever barns in this area (see figure 18).[20] The Robertson barn also has a post-and-lintel loft frame, as does the John A. Seaton barn, the only two barns in the district to have this structural configuration. All three barns seem to have been constructed in the mid-1870s, so their common characteristics are not hard to explain. Unfortunately, there is nothing available to explain why the Robertson barn is a half-double cantilever.

The other two half-double-cantilever barns on Roberts Schoolhouse Road were built about 1885 by brothers, Sevier 165 by William M. Roberts (1854-1923) and Sevier 166 by Stephen H. Roberts (1855-1907).[21] Their sister, Martha Roberts, was the wife of John A. Seaton, and their barns are located on the same lane off Roberts Schoolhouse Road. One might expect these barns to be similar—they are of the same general type and scale, and they were erected at about the same time—but few dimensions in one barn correspond with the same dimensions in the other, causing one to wonder if similarities in proportion observed between other barns are coincidence or evidence of influence from one to the other.

The second-largest cantilever barn on Roberts Schoolhouse Road, Sevier 22, was built about 1875 by William W. Webb (1834-1908), husband of William Phillip Roberts's eldest daughter, Alvira Roberts.[22] Webb purchased the farm of Anderson Seaton Jr. in 1873 and built on it a barn with fourteen secondary cantilevers like the contemporary James H. Seaton barn. With a family of ten children, he would have required a considerable farm and barn.

Three of the double-cantilever barns built along Roberts Schoolhouse Road have remarkably similar proportions and dates (see figure 52). John Ogle (1848-1930) built Sevier 172 about 1880.[23] Ogle was the grandson of Horatio Butler, and his farm was located close to the intersection of Roberts Schoolhouse and Middle Creek roads. Its cribs are the same size as those of Sevier 152, built about 1885 by Jacob H. Seaton (1856-1909), younger brother of James H. Seaton.[24] This Seaton barn, in turn, is virtually identical in dimensions and features to Sevier 156, built by Henry Butler (1826-1916) on Shinbone Road, at the far end of Roberts Schoolhouse Road.[25] Could these similarities be a matter of coincidence? Beulah Linn forwards an interesting interpretation, that all might be the work of one carpenter, William J. Trotter, a great-grandson of both John Trotter of Sevier 5 and John Roberts, whose son built Sevier 173.[26] The granddaughter of Henry Butler, Mattie Belle Butler Householder, has remarked that "all the old barns on Middle Creek and Roberts School Road are Trotter barns." Beulah Linn notes,

> The house I live in was built 1887. The chief carpenter was a William Trotter. When I moved here (1954) I had a huge cantilever barn torn down as the roof was falling and I didn't have the money to fix up the house and repair the barn, too. However, my great-great grandfather had settled on the land in 1824, so I believe the barn was here before the present house was built in 1887. The Wm. Trotter who helped my great-uncle Henry Gobble build the house is listed as a carpenter, age 28, in the 1880 census.[27]

The Butler barn is inscribed "H. Tedford 1888" on the front of one crib log, which might be understood as the signature of a builder. Mattie Householder provided the answer: H. Tedford was her uncle who lived in Blount County, and he probably scratched his name on the barn while on a visit. His writing may indicate that the barn was built before 1888.

The last cantilever barn along Roberts Schoolhouse Road, Sevier 139, belonged to William M. Sutton (1866-1941), who married Ellen Roberts, a granddaughter of William Phillip Roberts, and probably built the barn about 1905.[28] The Sutton barn is now partially destroyed, the crib logs having been removed and replaced by posts which continue to support the secondary cantilever beams; and the log house which belongs with the barn has been given exterior clapboarding. Otherwise, the farm presents a picture of late nineteenth-century rural life: set slightly away from the main road, house and barn face each other across a small stream.

The Middle Creek community contained other families, of course, but none whose cantilever barns survive to the present. (It is impossible to say who did *not* have a cantilever barn.) One rather prosperous Middle Creek farmer of interest because of his family name was Robert Myomey Rambo (1829-1908). *Rambo* is one of the names identified by Jordan and Kaups as being traceable to the Finno-Swedish settlements along the Delaware, where horizontal log construction was first practiced in the New World. Could he perhaps provide a link to Old World building practices? Rambo's parents were not from Tennessee: his father, Peter Rambo (1787-1846), was born in Virginia and received a land grant in Tennessee after service in the War of 1812, while his mother, Mary Frances Marshall (1798-1873), was born in Pennsylvania. Although the Peter Rambo farm was located

about halfway from Sevierville to Pigeon Forge, the Marshall family home was along Middle Creek (south of the Roberts Schoolhouse Road intersection), and it was there that Peter and Frances were married in November of 1814. Their son, Robert M. Rambo, served in the Confederate Army and married Lyda A. Emert (1836-1913), eventually settling at his mother's family home. Older residents of the community recall the Marshall home (now gone), but there is no memory of a cantilever barn on the property.[29] Another interesting speculation thus remains unresolvable.

Observations

From this study of the Middle Creek cantilever barns, it is evident that many families in the community were related by blood or marriage, and it seems likely that barn building was influenced by these family relationships. There is also the strong possibility that carpenters (who perhaps were also relatives) may have been involved with the construction of cantilever barns, especially in the period after 1880, thus accounting for some common features, proportions, and minor details that are seen in localized settings. If what happened in Middle Creek is typical of the larger area, then it is easy to imagine that this story repeated itself across the county and into neighboring Blount County while the tradition of building cantilever barns did not spread significantly beyond this rather limited region. Farms were most likely to have a cantilever barn if neighboring farms also had one or if the farmer were related to the owner of one. The network of personal or family contacts kept the practice of cantilevered construction within a small area.

CHAPTER 6

THE BARNS AS VERNACULAR EXPRESSION

Popularity proved to be a transient attribute of cantilever barns. Once so useful, these barns have by the late twentieth century become vestigial elements of a vanishing agricultural past. Reasons for their decline are not difficult to find. Studies of Sevier County conducted by the Smoky Mountain Historical Society have documented the gradual changes after 1918 which brought a remote nineteenth-century agricultural county into the twentieth century. Increased contact with the outside world made possible by improved roads and new technology—automobiles, radio, and electricity—and accelerated by participation in the First World War contributed to modernizing trends that made it less likely that traditional folkways would be maintained.[1]

Furthermore, farming practices advocated on the basis of agricultural research recommended the building of specialized farm structures according to the particular animal to be housed or the materials to be stored: a sty for pigs, a pen for turkeys, a garage for tractors, a dairy barn for cows, a silo for silage, and so forth. Multipurpose barns such as those of the cantilever variety were not encouraged, and while few may have been demolished to conform

with new ideas, the construction of new ones may have been curtailed. A number of cantilever barns have survived to the present by themselves becoming single-purpose structures, useful primarily for drying burley tobacco, which became a major cash crop as transportation improvements of the twentieth century made markets accessible. Of those barns still in use as farm structures, the majority are serving much as they did when built: animals—most often cattle—are housed in the cribs, and hay is stored in the loft. The great cantilever timbers have been able to support the loads of modern rectangular bales, which have a greater density than loosely forked hay. Whatever their other functions, fully a third of all cantilever barns are hung with tobacco each fall.

None have survived without modifications in use. Most commonly, this has involved the renewal of exterior cladding, the original horizontal lapped siding being replaced generally by roughly sawed vertical boarding and the original wood shingles being replaced universally by galvanized metal sheet roofing. Complete enclosure of the log structure, either by extending siding

Figure 53. *Two-crib double-cantilever barn masked by additions and enclosures (Sevier 112).*

to the ground or by the addition of lean-to sheds along one or more of the barn's sides, has happened to fully 80 percent of surviving barns, thereby disguising their cantilevers (figure 53). Only by going inside can the original logwork still be seen. Many cantilever beams now have supported ends, some for

obviously needed structural reinforcement and others because these additions facilitate the installation of doors and siding.

A final modernizing trend that has indirectly affected cantilever barns was the advent of the TVa in the 1930s. TVA brought not only affordable electricity to rural Tennessee but also a commitment to land conservation. Steep hillsides were reforested, and year-round pastures were recommended for gentler slopes as a means of holding the soil. Farmers were urged to refrain from growing seasonal crops on erosion-prone lands and to shift away from corn to beef or dairy cattle farming. Cantilever barns could not accommodate more than a few animals in each crib, making them less useful for larger herds, and the unsanitary conditions of their dirt floors rendered them wholly unsuitable for dairy farming. Good practice requires that dairy herds not share barn space with any other livestock. In Blount County, where the shift to dairy or cattle farming has been more extensive than in Sevier County, there are fewer cantilever barns today, perhaps as a result of dairy farming.

Studies of known barn builders would seem to confirm that the earlier barns are generally the largest and most substantial. Most commonly, the farmers who built the early barns were more prosperous than average, and several early builders were also men of local social distinction. Their barns may well have served as models for imitation by others. Farmers appear to have built barns sized to the relative prosperity of the farm; farm value in 1880 and barn size are correlated among owners of double-cantilever barns. By the 1920s and 1930s, barn size and construction quality show a marked decline from the earlier structures.

The general but local popularity of cantilever barns has parallels in other barn studies which have demonstrated that these buildings are often quite regional, contributing greatly to the character of rural landscapes. The diverse agricultural areas of the United States have presented many opportunities for the development of varied farm buildings, and these traditions deviate widely as one surveys the climate, topography, farm

animals, and crops raised across the country. In response to these differing conditions, barn designs in many parts of the United States are often highly localized. Regional cultural backgrounds and predilections, available materials, and patterns of agriculture also impact rural architecture, and nowhere is this more the case than with barns. In New England, for example, the custom of linking the house and barn with a series of smaller outbuildings has created a distinctive vernacular tradition. Research on these barns reveals that far from being accidental, the connection of farm buildings was actively encouraged in New England by state agricultural agents (the forerunners of the United States Agricultural Extension Service) through the agricultural press. Farmers responded by constructing new barns or moving existing ones to align with passages from the main farmhouse.[2] Similar ideas of "progressive" farming spurred the construction in various midwestern states of circular or polygonal barns, a form first advanced by George Washington at Mount Vernon and later used by the Shakers in New England. In the era when octagonal plans were widely promoted for efficient and healthful homes, the same form was advocated for barns for similar reasons: a large area was encompassed by the least amount of perimeter wall, and the central raised roof area provided effective air circulation on the interior.[3]

In contrast to these barn forms, whose origins and dispersal can be largely explained through regional or national farming publications, the cantilever barn appears to have developed without any assistance from the written word. Just as there are no contemporary manuals explaining how to build a log cabin, there seems to be no recorded explanation of cantilever barn construction. These barns were constructed in an area where crib-barn plans were already established, where knowledge of both log and heavy timber frame construction was broadly shared, and where availability of material made these barns affordable to a general spectrum of farmers. As has been shown, the combination of family connections, geographic proximity, and the social position of early barn owners probably accounts for their general adoption. It might be safe to claim that in the period when log barns were dominant in Blount and Sevier counties, the vast majority of them were of cantilevered construction.

The occurrence of cantilever barns in East Tennessee and almost nowhere else is interpreted as a local invention, developed in the same spirit that inspired others elsewhere to explore circular plans, "doughnut" or oval barns in Illinois,[4] or the connected farm buildings of New England, all of which testify to the inventiveness of American farmers in the design of barns. In this context, it should not be a surprise to find that a close-knit farming region originated a distinctive architectural solution in response to local conditions. Cantilever barns were derived in part from established prototypes—including the Pennsylvania German barn, German two-crib barns, and log blockhouses—but they have sufficient original features to justify the label *innovation*. The idea did not spread widely, because other areas had developed other solutions, making it unnecessary to import designs pioneered elsewhere. Cantilevered construction was one possible response to a particular set of conditions, but it was not the only way (or necessarily the best way) in which wooden barns could be built. The form worked well enough to be widely imitated within a limited region, however, and as farmers are prone to conservative and traditional ways, this locally developed form persisted.

How do cantilever barns fit into the larger world of American barn forms? Our study leads us to concur with Allen Noble, who has diagrammed the two-crib double-cantilever form as a special case of two-crib barns, having no "offspring" in the developmental chart (see figure 27). The picture becomes less clear when the remaining four types of cantilever barn are considered. One remarkable aspect of East Tennessee barns is the variety of basic barn plans to which the cantilever principle was applied: single-crib, two-crib, and four-crib types were all used. The desire to build using cantilevered construction was apparently stronger than allegiance to any one particular form, and this quality, rather than any specific plan type, is the most important aspect of the barns. To elaborate upon Noble's schematic diagram, one might modify the chart to note that cantilevers are found on three different crib plan arrangements (figure 54).

This account of the cantilever barn leaves unanswered many questions concerning agricultural society in East Tennessee during the nineteenth century. Its economic life, cultural pattern, and even its physical appearance are incompletely sketched through a study of surviving cantilever barns. Nevertheless, this project has documented more cantilever barns than will ever be available again, and it is hoped that this work will advance future investigation of the rural mosaic of which cantilever barns were once a part.

It is also hoped that this effort will contribute to the preservation of these fascinating cultural artifacts. In one sense, too many barns remain for people to become concerned over the disappearance of a few; but it is also true that, aside from the three barns located within the Great Smoky Mountains National Park (Blount 8, Blount 13, and Sevier 24) and one Sevier County barn (Sevier 183) moved to the Museum of Appalachia in Anderson County, no other cantilever barns are certain of continuation into the twenty-first century. There are some splendid specimens which ought to survive as evidence of nineteenth-century rural life in East Tennessee. Over the interval from 1984

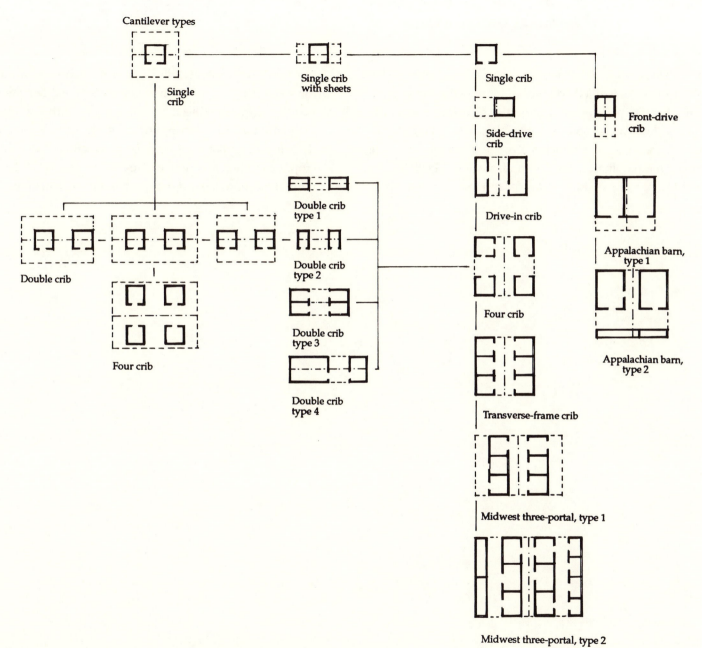

Figure 54. *Chart illustrating the position of cantilever barns within the category of crib barns, revised from the original scheme presented by Allen Noble as figure 24.*

to 1991, at least nine of the barns inventoried are known to have burned, been dismantled and moved, or blown down in windstorms. Far too many others are simply collapsing in place through neglect as farming becomes a marginal activity, giving way to vacation homes, tourist attractions, or the suburban extensions of growing metropolitan areas. The land, always difficult to farm, now has a higher real estate value for residential or commercial use than it does for agriculture, so understandably farming is on the wane. Most barns remaining in sound condition are that way thanks to tobacco, still a major money crop, although one whose long-term outlook is uncertain. Ironically the preservation in use of these ingenious barns depends on production of an unhealthy crop, a condition that underscores their precarious future. These reminders of our agricultural past deserve a better fate.

APPENDIX A

AGRICULTURAL GRAPHS

The vertical line represents the county average for each item on the graph.

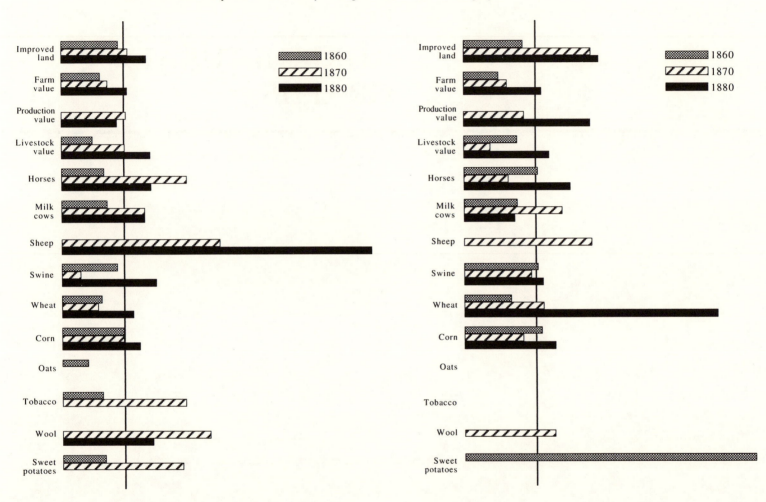

Figure A1. *Agricultural census data for the farm belonging to John Marshall, 1860-80. The vertical line represents the county average for each item on the graph.*

Figure A2. *Agricultural census data for the farm belonging to Robert Marshall, 1860-80. The vertical line represents the county average for each item on the graph.*

The vertical line represents the county average for each item on the graph.

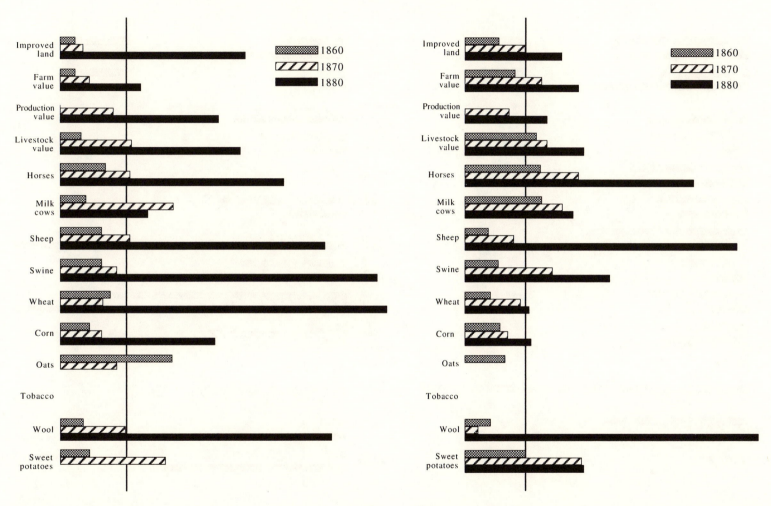

Figure A3. *Agricultural census data for the farm belonging to William W. Webb, 1860-80. The vertical line represents the county average for each item on the graph.*

Figure A4. *Agricultural census data for the farm belonging to Pryor L. Duggan, 1860-80. The vertical line represents the county average for each item on the graph.*

The vertical line represents the county average for each item on the graph.

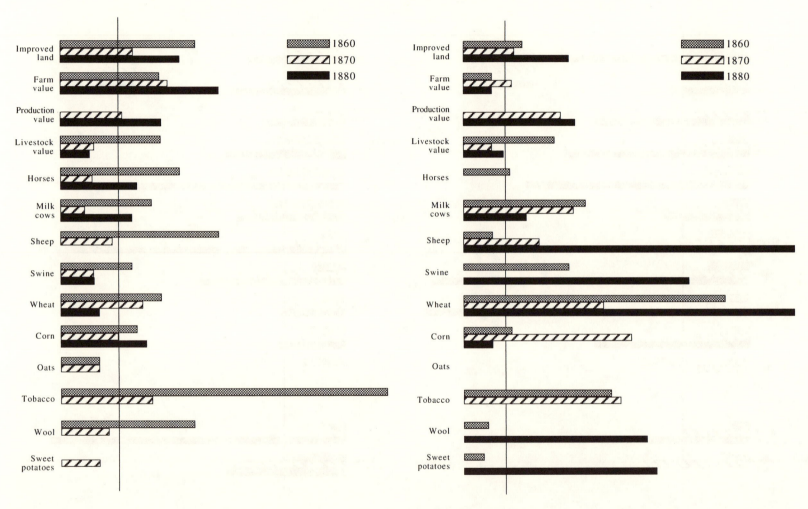

Figure A5. *Agricultural census data for the farm belonging to John Andes, 1860-80. The vertical line represents the county average for each item on the graph.*

Figure A6. *Agricultural census data for the farm belonging to Richard Reagan, 1860-80. The vertical line represents the county average for each item on the graph.*

The vertical line represents the county average for each item on the graph.

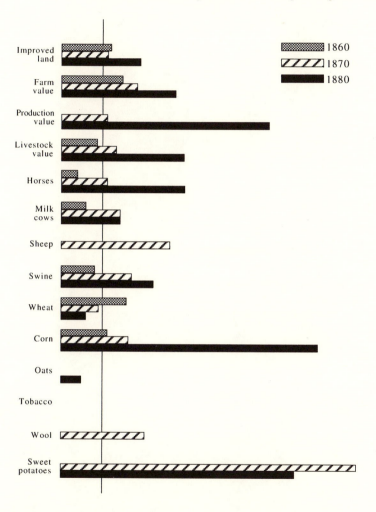

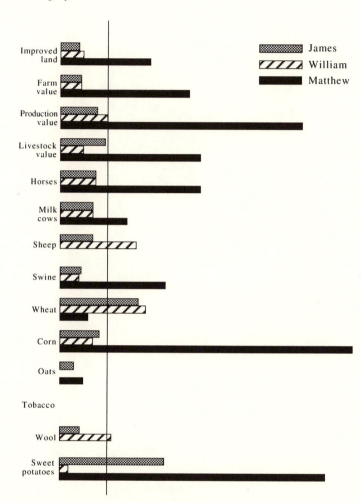

Figure A7. *Agricultural census data for the farm belonging to Matthew Tarwater, 1860-80. The vertical line represents the county average for each item on the graph.*

Figure A8. *Agricultural census data for the Tarwater family farms, 1880. The vertical line represents the county average for each item on the graph.*

APPENDIX B

BARN LOCATIONS AND DIMENSIONS

Included in this appendix are charts and detailed maps showing the locations of all cantilever barns. The maps are based on those provided by the Tennessee Department of Transportation. Each barn's location is indicated on the map by a dot and a number corresponding to the sequence number for the given county. In the charts giving measurements of all barns, the barns are arranged by types. Within each type, they are listed alphabetically by county, and the number on the map is preceded by a two-letter abbreviation for the county. Thus Sevier County becomes SE, Knox County is KN, and Blount County is BL.

All measurements were made using English units. Building dimensions are given in feet and inches, while timber sizes are expressed in inches (width x depth). The symbol ø indicates a round log cross-section. Blanks in columns mean that a particular members are missing or could not be measured. The abbreviated columns presented after the dimensions record construction materials, details, and information regarding condition, orientation, and use.

The original field data sheets, maps, and all correspondence related to this project have been deposited in the Special Collections Library of the University of Tennessee, Knoxville.

L = number of crib logs
N = notch type (H for half-dovetail, S for square, V for V-notching)
T = tool marks observed, generally Ax, Az (adz), or Saw (circular saw)
F = frame type, either P (post and lintel) or Q (modified queen post)
S = siding, generally H (horizontal boarding), V (vertical), or Al (aluminum)
R = roof material, usually M (galvanized sheet metal) or W (wood shingle)
C = condition, denoted as G (good, sound condition), F (fair; needs repair), or P (poor, in danger of collapse)

Use of both the crib and loft is indicated by the following letters:
A = animals (various); C = cows; Ch = chickens; E = empty;
    H = hay; Ho = horses; P = pigs; St = storage; T = tobacco;
    Tu = turkeys
O = orientation of the barn, generally the direction of crib openings
Notes give the builder's dates (if available) and other miscellaneous information. "Extensions" refers to the length of inner extensions of the primary beam.

Table B1
Information Tables for All Barns Surveyed

Two-Crib Double-Cantilever Barns

	BUILDER	LOCATION	OVERALL DIMENSIONS		WIDTH		DEPTH				CANTILEVER Primary	
			width	depth	cant	crib	space	cant	crib	#	end	mid
BL1	Joe L. Delozier	U.S. 411/River Rd.	70-0	37-9	10-0	15-0	20-0	10	17-9	16	8x8	8x14
BL2	Carl Trundle	Porter School Rd.	56-0	30-11	7-10	12-2	20-0	7-10	15-3	16	7x7	7x11
BL3	Andy M. Delozier	Green Rd.	73-1	42-10	10-4	17-9	17-0	8-11	25-0	16	9x9	10x12
BL6	John H. Ogle	W. Miller's Cove Rd.	49-5	30-6	7-9	11-10	10-3	8-3	14-0	8	9x8	9x15
BL7	W. R. Taylor	Cameron Rd.	43-2	24-10	6-3	10-4	10-0	6-5	12-0	8	7x6	7x9
BL8	Hamp Tipton	[Cades Cove Rd.]	45-0	29-9	6-7	11-11	10-4	9-0	11-9	11	8x8	8x13
BL10	Nathan Myers	Tuckaleechee Rd.	48-0	33-9	6-5	12-2	12-0	9-0	15-8	8	8x9	8x12
BL11	Mack Davis	Walter Davis Rd.	64-3	34-8	10-3	11-11	20-0	8-4	18-0	18	10x10	10x14
BL12	Ben Rogers	H. R. Davis Rd.	51-5	31-9	8-0	11-9	11-11	8-11	13-11	12	6x8	6x12
BL13	Anthony	[Cades Cove Rd.]	55-8	30-5	9-0	12-7	12-6	9-0	12-5	8	8x6	8x12
BL14	McNeilly	U.S. 321	62-8	38-2	11-0	12-0	16-8	10-2	17-10	12	9x8	11x13
BL15	James B. Bogle	Jefferies Hollow Rd.	58-4	31-10	8-2	14-0	14-0	7-11	16-0	14	11x12	11x18
BL16	Will Davis	Jefferies Hollow Rd.	62-0	34-0	12-0	12-0	14-0	8-0	18-0	16	10x12	10x12
BL17	Jake Harmon	Boling Rd.	49-10	20-0	6-11	12-0	12-0	4-0	12-0	10	8x7	9x8
BL18	Jack Graves	H. R. Davis Rd.	53-8	33-9	7-1	11-11	15-8	9-0	15-9	10	7x9	8x12
BL19	Steve P. Graves	Chilhowee Rd.	56-3	34-2	8-0	12-0	16-3	9-0	16-2	12	6x8	9x13
BL20	Nelson Garner	Chilhowee Rd.	46-5	25-0	7-10	9-6	11-9	5-6	14-0	10	8x7	9x13
BL22	Eli Garner	Chilhowee Rd.	52-7	35-2	7-6	12-0	13-7	8-6	8-2	12	7x9	7x12
BL23	Samuel M. Kounts	McKendry Rd.	56-2	32-5	8-0	12-0	16-2	8-1	16-3	12	9x11	9x15
BL24	Andy Davis	Jim Norton Rd.	62-8	36-0	10-2	12-0	18-4	9-0	18-0	12	9x7	8x11
BL25	Reid Allen	Chilhowee Rd.	39-4	23-10	4-3	10-5	10-0	5-11	12-0	10	7x6	7x10
BL26	Whitehead	Blair Rd.	46-11	30-0	6-0	11-9	12-5	8-0	14	10	8x10	9x11
BL27	———	U.S. 129	55-8	33-6	9-11	12-0	11-10	9-8	14-2	14	8x8	8x12
BL28	———	W. Miller's Cove Rd.	63-8	30-0	8-0	11-10	14-0	8-0	14-0	10	9x15	9x15

| DIMENSIONS | | BARN HEIGHTS | | | | | | | | | | | USE | | | |
| Secondary | | | | | | | | | | | | | | | | |
end	mid	crib	eave	ridge	L	N	T	F	S	R	C	crib	loft	O	Notes
8x8	8x14	6-7	6-9	8-4	5	H	Ax	P	H	M	F	C	H		Extensions 5-0. JLD 1896–1954.
8x7	7x14	7-0	7-6	10-6	7	H	Ax	Q	H	M	F	C	H	NE	CT 1894–1956.
9x9	7x14	7-8	5-10	—	6	B	Ax	Q	H	M	G	C	H	NE	Exts. 2-11. AMD 1864–1939.
7x8	8x13	7-4	7-6	—.	4	H	Ax	Q	H	M	F	Ho	H	NE	Center doors. JHO 1893–1973.
7x6	7x10	7-0	5-0	5-0	6	H	Ax	P	H	M	P	A	H	NW	
9x8	8x16	6-8	4-0	9-4	5	V	Ax	P	V	W	G	E	E	SE	1968 reconstruction in GSMNP.
9x9	10x14	6-6	5-0	15-0	5	H	Ax	P	V	M	F	St	E	N	Built from parts of other barns.
7x8	7x8	5-4	6-8	16-6	5	H	Ax	P	H	M	F	E	T	NE	Mutilated. MD 1867–1936.
5x6	7x9	6-4	6-0	12-6	5	H	Ax	Q	H	M	P	E	E	SW	"1893" carved on N. primary.
9x6	9x12	6-2	5-4	7-6	4	S	Ax	Q	V	W	G	E	E	SE	Moved to GSMNP. Pre-1890.
10x7	9x13	5-6	8-11	18-0	4	H	Ax	Q	H	M	G	St	H	NE	Extensions 3-0. c. 1880.
10x8	10x15	6-2	5-0	13-6	5	H	Ax	—	H	M	P	H	—	N	Extensions 2-0. JBB 1856–1933.
8x10	8x10	6-3	8-0	16-0	4	H	Saw	Q	H	M	G	C	E	S	c. 1900.
6x6	6x6	6-4	5-0	7-6	7	H	Ax	P	V	M	F	C	H	SE	1900–1910.
7x9	7x12	6-8	6-3	12-6	4	H	Ax	Q	H	M	F	C	E	N	Extensions 3-0. c. 1875.
6x9	6x16	6-0	5-0	10-0	4	H	Ax	Q	H	M	F	E	H	SE	Extensions 3-3. SPG 1870–1947.
6x9	7x10	5-2	5-8	9-0	4	S	Ax	P	H	M	P	St	T	NE	1890s. NG 1863–1925.
7x8	8x12	6-6	5-0	9-6	3	H	Ax	P	H	M	P	A	A	S	c. 1865. EG 1814–1900.
9x11	9x14	6-6	7-4	—	5	H	Ax	—	H	M	G	E	H	SE	Extensions 3-1. SMK 1836–1911.
8x6	8x12	5-0	5-0	6-6	5	H	Ax	P	H	M	P	P	E	SE	Extensions 4-0. Pre-1900.
8x7	8x12	5-6	4-4	7-0	5	V	Ax	P	Al	M	G	E	T	SW	Pre-1910.
9x9	9x18	7-7	6-3	10-0	5	H	Ax	P	H	M	F	St	T	NE	Built of yellow pine.
8x8	8x10	5-7	7-0	10-6	5	H	Ax	Q	H	M	P	C	T	NW	Pre-1900.
9x8	9x13	7-5	4-6	10-0	5	S	Ax	Q	H	M	P	H	T	W	Threshing floor in center.

Table B1 continued
Information Tables for All Barns Surveyed

Two-Crib Double-Cantilever Barns

	BUILDER	LOCATION	OVERALL DIMENSIONS width	depth	WIDTH cant	crib	DEPTH space	cant	crib	#	CANTILEVER Primary end	mid
BL29	John Stephenson	Schular Rd.	50-1	30-0	8-0	12-2	9-9	8-0	14-0	—	—	—
BL30	William Ross	Big Springs Rd.	62-0	37-4	10-0	14-0	14-0	9-8	18-0	6	8x12	7x15
BL34 1974.	Samuel J. Cook	Dunlap Hollow Rd.	51-9	39-9	—	—	—	—	—	—	—	—
BL35	Cam Walker	Marble Hill Rd.	60-4	36-8	10-0	11-6	17-4	9-10	17-0	12	7x8	7x12
BL36	Kaiser	Marble Hill Rd.	62-3	35-10	10-3	12-3	17-6	9-8	16-6	8	8x7	8x13
BL37	———	Marble Hill Rd.	61-11	40-4	12-2	11-8	14-3	10-2	20-0	10	6x6	6x10
BL38	Joseph L. Jones	Morganton Rd.	54-7	40-0	10-2	14-0	12-3	10-0	20-0	12	10x11	11x12
BL46	Bart Dudley	Doc Norton Rd.	48-7	32-3	5-0	11-0	16-7	8-0	16-3	10	7x10	7x14
BL47	Henry E. Sawyer	Prospect Rd.	53-3	32-8	9-0	12-0	12-3	8-4	16-0	10	7x12	7x12
BL49	Bogle	Prospect Rd.	51-11	31-10	7-10	12-0	12-3	8-0	15-10	10	9x9	9x14
BL50	Clyde D. Wilson	Jim Norton Rd.	52-0	31-10	8-0	12-0	12-0	8-0	15-10	10	8x8	8x13
BL51	John F. Tipton	Ellejoy Rd.	54-5	29-5	7-6	12-5	14-7	7-6	14-5	10	7x10	8x16
BL56	Jesse Brown	Brown School Rd.	56-9	36-0	9-10	12-3	16-2	10-0	16-0	11	8x8	8x12
BL57	Keller	Brown School Rd.	66-8	40-0	10-5	13-0	19-10	9-11	20-2	12	9x9	9x12
BL58	Samuel Rorex	U.S. 411	61-6	38-6	8-7	14-1	18-2	10-4	17-10	14	11x9	11x10
BL59	Enoch Waters	Hitch Rd.	60-1	36-0	10-0	11-11	18-3	10-0	16-0	12	9x9	9x9
BL62	Eli Farmer	Cold Springs Rd.	54-0	35-11	9-0	11-10	12-4	11-4	13-3	12	—	—
BL65	———	Roddy Branch Rd.	59-5	39-1	8-6	12-0	18-5	9-0	21-1	10	7x12	7x17
BL66	W. R. Taylor	Carr Creek Rd.	41-10	32-0	6-0	9-10	10-2	8-0	16-0	8	7x10	7x17
BL74	George W. Ross	Grand View Rd.	48-4	30-0	8-0	10-0	12-4	9-0	12-0	10	6x6	6x8
BL75	John H. Coker	Ralston Rd.	62-0	36-0	10-0	12-0	18-0	10-0	16-0	14	8x6	8x12
BL76	John H. Blevins	Old Piney Rd.	58-10	36-1	9-0	11-10	17-2	9-0	18-1	12	9x7	9x9
BL77	Tom Keller	Galyon Rd.	53-7	28-0	8-0	13-10	9-11	8-0	12-0	8	8x12	8x12
BL78	Samuel W. Keller	Galyon Rd.	49-9	30-3	7-8	12-0	10-5	7-0	16-3	8	8x9	8x11

DIMENSIONS Secondary		BARN HEIGHTS										USE			
end	mid	crib	eave	ridge	L	N	T	F	S	R	C	crib	loft	O	Notes
6x6	6x8	7-6	—	—	—	—	—	—	—	—	—	E	E	SE	Cantilevers gone. JES 1847–1937.
8x8	8x15	7-3	6-0	10-0	5	H	Ax	P	—	M	P	St	E	NW	Pre-1870. WR 1810–67.
—	7x5	—	—	—	—	V	Ax	Q	H	M	F	St	St	NW	Reverse cantilever. SJC 1893–
8x7	8x7	6-6	6-10	8-6	7	V	Ax	Q	H	M	G	P	H	SE	Reverse cantilever. 1890s.
8x6	8x6	7-4	7-9	8-6	6	V	Ax	Q	H	M	P	E	T	NW	Reverse cantilever. Pre-1865.
6x5	6x10	6-8	6-8	12-6	6	H	Ax	Q	H	M	F	C	T	SW	Pre-1900.
8x7	8x11	5-7	8-0	15-0	5	H	Ax	Q	H	M	F	C	H	SW	Pre-1900. JLJ 1873–1943.
8x7	8x7	8-0	5-5	8-4	5	H	Ax	Q	H	M	G	St	H	SE	1932.
7x9	7x9	6-3	6-10	15-0	4	H	A I S	Q	H	M	G	E	E	NW	1-0 extensions. HES 1891–n.d.
7x9	7x11	7-5	6-0	12-6	5	H	Ax	Q	H	M	F	C	H	E	Pre-1890.
8x8	8x10	6-10	6-0	9-0	5	H	Ax	Q	V	M	G	C	H	NE	1890s. CDW 1881–1919.
8x6	8x11	6-5	5-4	9-0	5	H	Ax	Q	H	M	P	C	H	W	Pre-1875. JFT 1855–1901.
7x7	7x10	6-0	6-10	14-6	—	—	Ax	Q	H	M	P	H	H	NE	1890s. Dissimilar cribs gone.
10x9	10x15	7-0	8-0	18-0	5	V	Ax	P	M	M	G	C	H	NE	5-1 extns. 3 primaries. Pre-1880.
7x10	7x10	7-0	9-10	16-8	5	H	Ax	Q	H	M	G	E	H	S	Cont. primary. SR 1843–1906.
7x9	7x13	6-9	6-10	11-8	5	S	Ax	Q	H	M	F	C	HT	SE	4-0 extensions. EW 1856–1942.
8x8	7x14	—	8-0	15-0	—	—	—	—	H	M	F	HT	HT	W	Pre-1900. Cribs gone.
11x10	11x10	6-8	7-0	10-0	6	V	Ax	Q	—	W	F	—	—	—	Used as a house. EF 1843–1912.
7x11	8x13	5-6	6-0	12-6	5	H	Ax	Q	H	M	G	St	T	NW	1880s. Poplar logs.
7x7	9x7	5-7	6-8	12-0	6	S	Ax	Q	V	M	G	H	T	—	c. 1910. GRW 1863–1939.
9x6	9x12	8-6	6-0	11-3	6	V	Ax	P	H	M	P	E	E	—	Cont. primary. JHC 1876–1939.
8x8	8x10	5-7	6-8	16-0	4	V	Ax	Q	H	M	G	Ho	H	—	Rt. crib ext. 5-0. JHB 1869–1923.
8x6	8x6	5-8	6-2	6-0	5	H	Ax	P	V	M	F	C	H	—	Cribs open to center. 1877.
7x7	7x11	6-8	6-0	7-6	6	V	Ax	Q	V	M	G	C	HT	—	Pre-1900. SWK 1882–1955.

Table B1 continued
Information Tables for All Barns Surveyed

Two-Crib Double-Cantilever Barns

	BUILDER	LOCATION	OVERALL DIMENSIONS		WIDTH		DEPTH				CANTILEVER Primary	
			width	depth	cant	crib	space	cant	crib	#	end	mid
BL79	Fate Hall	Mel Hall Rd.	60-0	36-2	10-0	12-0	16-0	10-0	16-2	6	7x7	7x11
BL81	James N.Grindstaff	Blockhouse Rd.	56-0	31-11	8-0	12-0	16-0	8-0	15-11	6	6x8	6x11
BL82	Samuel H. James	James Rd.	45-7	32-0	4-0	12-2	13-3	9-0	14-0	8	7x11	7x11
BL84	John G. Coulter	Old Walland Highway	—	30-9	8-0	12-0	—	8-2	14-5	12	8x8	9x12
BL86	Morton	Lee Lambert Rd.	61-2	37-10	11-0	11-10	16-6	10-0	17-10	12	9x9	10x13
BL88	George A. King	Lee Shirley Rd.	50-7	34-2	7-9	11-10	11-5	9-9	13-8	6	5x11	5x11
BL90	George Ross	Lanier Rd.	56-0	34-2	9-9	12-3	12-0	10-4	13-6	14	7x11	7x11
BL95	Bingham	Bingham Lane	56-0	33-0	10-0	12-0	12-0	8-6	16-0	13	9x9	9x7
BL100	John Long	Six Mile Rd.	38-2	28-0	4-3	7-9	10-6	8-2	11-8	6	9x8	—
BL101	James Taylor	Mint Rd.	50-2	41-0	10-3	11-10	16-0	10-6	20-0	14	8x7	8x10
BL102	Carter Stout	Six Mile Rd.	51-9	32-2	8-0	11-10	12-1	9-0	14-2	10	8x7	8x7
CA1	———	TN 91 at Winner	67-8	35-5	10-0	15-8	16-4	9-8	16-0	6	—	—
GR1	Bright	TN 93/Union Temple	68-9	42-5	11-2	12-8	20-1	12-0	18-5	6	7x9	—
GR2	Moore (?)	Carl Doty Rd.	79-4	43-1	11-0	19-8	18-0	11-8	19-9	6	8x12	—
GR3	———	Rheatown Rd.	49-0	40-2	9-6	12-0	6-0	10-0	30-2	6	7x7	7x10
JE2	Thompson	Spring Creek Rd.	60-3	40-9	7-3	15-10	14-0	10-6	19-9	8	9x9	9x9
JO2	Bessie Wilson	Stouts Branch Rd.	53-11	36-6	9-7	12-6	9-9	10-1	16-4	4	7x10	8x13
JO4	———	TN 167 at Butler	56-10	31-8	10-3	12-9	10-10	9-6	12-7	6	7x6	7x6
JO7	———	Furnace Creek Rd.	70-7	37-4	11-3	17-0	14-1	10-8	16-0	—	8x15	8x13
KN1	Brakebill	Maloney Rd.	63-11	40-2	10-0	14-1	15-10	10-10	17-1	10	10x8	10x15
KN2	Peter Johnson	Crenshaw Rd.	50-0	38-7	8-6	12-3	10-1	9-5	19-9	10	10x8	10x14
KN3	Doyle	Martin Mill Pike	57-3	34-0	10-0	11-8	16-3	10-1	13-10	6	9x8	9x12

DIMENSIONS Secondary end	mid	BARN HEIGHTS crib	eave	ridge	L	N	T	F	S	R	C	USE crib	loft	O	Notes
7x8	7x17	5-6	6-2	11-3	4	S	Ax	Q	HV	M	G	St	H	—	c. 1885.
6x9	6x12	6-8	6-8	14-0	5	S	Ax	Q	V	M	G	C	HT	—	Sawed oak cribs. JNG 1863–1930.
8x8	8x10	5-7	6-0	8-0	5	SH	Ax	Q	V	M	F	H	H	—	Extensions 1-10. SHJ 1869–1964.
8x8	8x10	6-10	8-0	14-0	5	V	Ax	Q	H	M	P	E	E	—	Extension 3-0. JGC 1829–1904.
8x8	8x12	—	8-5	16-0	4	H	Ax	Q	H	M	F	C	H	—	Extension 3-0. c. 1880–85.
7x12	7x12	5-8	5-4	11-3	5	H	Ax	P	H	M	G	St	T	—	Center open cribs. GAK 1837–93.
6x11	6x11	6-3	8-0	11-0	5	H	Ax	Q	H	M	G	C	H	—	c. 1880.
12x6	12x8	6-6	—	—	6	V	Ax	—	H	M	P	E	E	—	Treshing floor. Pre-1900.
10x8	10x8	6-6	5-5	10-0	7	V	Ax	Q	HV	M	G	T	T	SE	c. 1900. CS 1863–1942.
12x8	12x8	6-3	7-0	16-0	5	H	Ax	Q	H	M	G	H	T	SW	Extensions 3-0. Pre-1900.
8x6	8x7	6-3	6-0	16-0	6	H	Ax	Q	H	M	G	H	H	S	Pre-1905.
8x11	8x11	8-0	6-0	12-0	6	V	Ax	—	V	M	F	St	T	NW	Two-story cribs.
7x8	—	6-10	5-0	10-0	6	V	Ax	—	V	M	F	C	HT	NW	Two-story cribs.
9x11	—	6-10	6-4	—	—	V	Ax	—	V	M	G	E	E	NE	Two-story cribs.
8x7	8x9	—	—	—	5	V	Ax	Q	V	M	P	C	H	NW	Three cribs. No center aisle.
—	10x7	6-7	—	—	5	H	Ax	P	V	M	G	Ho	HT	N	Extensively reworked.
6x10	8x14	5-4	3-9	8-2	4	H	Ax	—	V	M	G	H	H	NW	Two-story cribs.
7x8	7x8	5-6	5-6	11-0	5	S	Ax	P	H	M	F	H	H	S	
8x14	8x19	8-0	6-7	10-0	4	H	Ax	—	V	M	G	St	T	N	Crib of white pine.
8x8	8x16	7-10	5-11	11-6	5	H	Ax	Q	H	M	F	H	St	E	Removed.
9x8	9x16	7-8	5-4	10-0	6	H	Ax	Q	H	M	P	St	H	E	Log trough in crib.
9x8	10x13	6-5	7-3	11-5	5	H	Ax	Q	H	M	F	St	H	W	Moved across road.

Table B1 continued
Information Tables for All Barns Surveyed

Two-Crib Double-Cantilever Barns

	BUILDER	LOCATION	OVERALL DIMENSIONS width	depth	WIDTH cant	crib	DEPTH space	cant	crib	#	CANTILEVER Primary end	mid
KN4	———	Hedron Chapel Rd.	60-0	34-2	8-3	12-0	20-6	8-1	18-0	—	6x9	6x10
KN6	Johnson	Huffaker Ferry Rd.	58-0	34-8	8-10	12-0	16-4	9-1	16-6	8	11x12	12x12
ME1	G. W. Shiflett	Sugar Creek Rd.	67-8	44-6	10-0	13-10	20-4	10-3	20-0	6	8x12	8x15
ME2	H. C. Shiflett	Sugar Creek Rd.	63-10	37-2	8-0	13-5	21-0	8-7	20-0	6	8x11	9x18
SE2	G.W. McMahan	Old TN 66 North	62-0	38-0	10-0	12-0	18-0	10-0	18-0	14	7x10	7x13
SE3	John W. Andes	Old TN 66 North	63-4	37-8	10-6	12-10	17-2	9-10	18-0	10	8x8	9x15
SE5	John Trotter	Middle Creek Rd.	64-5	39-4	9-5	12-10	20-5	10-4	20-0	18	10x10	9x14
SE6	William H. Trotter	Middle Creek Rd.	53-2	31-10	9-0	12-0	11-2	9-0	13-10	14	8x8	7x12
SE7	Pryor L. Duggan	Valley Rd.	50-9	34-1	7-6	11-6	11-9	7-9	18-7	8	—	—
SE8	———	US 441 West	49-9	32-2	8-0	10-0	13-11	8-1	16-0	10	9x6	9x11
SE9	Mathew Tarwater	US 441 West	55-5	31-0	7-11	13-2	13-5	7-8	15-8	10	12x6	12x12
SE10	T.D.W. McMahan	Pittman Center Rd.	62-0	36-0	9-10	12-2	18-0	8-10	18-4	14	10x12	9x12
SE14	T.D.W. McMahan	Pearl Rd.	81-11	38-0	10-0	24-0	13-11	10-0	18-0	18	8x12	8x12
SE17	McMahan	Bird Creek Rd.	52-6	33-8	8-0	12-0	12-6	8-10	16-0	12	12x8	12x15
SE18	Riley H. Andes	Old TN 66 North	63-1	36-8	10-0	12-0	19-0	9-4	18-0	10	7x8	7x18
SE20	Isaac Trotter	Middle Creek Rd.	58-11	37-10	10-6	12-4	12-10	9-11	18-0	12	7x12	7x14
SE22	William W. Webb	Roberts Schoolhouse	56-3	36-0	10-0	13-0	10-3	10-0	16-0	14	9x12	9x12
SE23	James H. Seaton	Seaton Springs Rd.	60-4	39-10	10-2	14-10	10-4	10-0	19-10	14	6x8	6x10
SE24	George Messer	Porter's Flat Trail	53-4	34-1	7-0	14-0	12-0	6-9	20-7	13	8x8	8x9
SE25	Sam P. McMahan	US 411 East	63-10	38-3	10-0	14-0	16-0	10-0	18-3	10	8x8	8x12
SE26	Darius Robertson	Richardson Cove Rd.	57-6	35-0	10-6	12-1	12-5	10-6	14-0	10	7x9	7x9
SE29	R. J. Ingle	Ingle Hollow Rd.	58-4	35-6	9-2	12-0	16-0	9-9	16-0	12	10x8	10x13

DIMENSIONS Secondary		BARN HEIGHTS										USE			
end	mid	crib	eave	ridge	L	N	T	F	S	R	C	crib	loft	O	Notes
—	11x7	6-10	7-7	15-0	6	H	Ax	Q	H	M	G	C	H	—	
11x9	11x12	7-6	7-6	10-6	5	H	Ax	Q	H	M	F	St	St	N	Log trough in crib.
8x10	8x17	7-6	—	—	6	H	Ax	Q	V	M	G	St	H	SE	Two pens each crib. c. 1860.
9x8	9x20	6-7	4-10	8-0	4	H	Ax	Q	H	M	G	H	H	SE	Extensions 2-9. c. 1866.
7x12	7x12	7-0	10-6	13-9	4	H	Ax	Q	H	M	F	C	H	S	GWMc 1842–1927.
8x8	8x15	6-0	6-8	13-6	5	H	Ax	Q	H	M	G	H	H	S	JWA 1838–1919.
8x9	8x9	6-4	7-6	11-8	5	H	Ax	Q	H	M	G	St	St	SE	Extensions 5-0. JT 1777–1856.
7x7	7x9	6-6	8-6	12-0	5	H	Ax	Q	H	M	G	C	H	E	WHT 1814–87.
10x9	10x13	7-8	7-0	9-0	7	S	Ax	—	H	M	F	C	H	N	PLD 1829–1906.
8x5	6x13	6-0	7-6	11-3	6	H	Saw	Q	H	M	G	A	T	SE	
8x6	8x11	6-8	4-6	16-4	5	H	Ax	Q	H	M	G	St	E	S	1880s. MT 1822–1902.
7x7	8x12	6-6	7-0	15-6	5	H	Ax	Q	H	M	G	St	E	S	Sawed timber in loft.
7x7	7x10	7-0	7-6	14-0	5	H	Ax	Q	H	M	G	C	H	N	TDWMc 1849–1921.
7x7	8x12	9-4	7-6	—	5	H	Ax	Q	H	M	G	A	H	E	
6x7	8x13	4-0	7-8	13-0	4	H	Ax	Q	H	M	G	C	H	NW	RHA 1835–1914.
7x7	7x13	8-4	5-7	11-3	5	H	Ax	Q	H	M	G	St	T	SE	Yellow pine. IT 1821–70.
6x6	7x11	7-4	8-6	12-6	5	H	Ax	Q	H	M	F	C	T	NE	WWW 1834–1908.
7x8	6x13	7-3	7-8	23-9	5	S	Ax	Q	H	M	G	Ho	T	NE	Center doors. JHS 1844–1916.
7x6	7x11	6-0	4-6	9-0	5	V	Ax	P	H	W	G	E	E	SE	Restored in GSMNP. GM b≈1843.
8x8	8x16	4-9	7-6	16-8	4	S	Ax	Q	H	M	P	St	H	SE	SPMc 1888–1960.
8x8	8x8	6-10	7-0	16-8	6	H	Ax	P	H	M	F	E	T	N	DR 1834–88.
9x5	9x10	7-9	5-3	12-2	5	H	Ax	Q	H	M	P	E	E	SW	4'3" extensions. RJI 1875–1966.

Table B1 continued
Information Tables for All Barns Surveyed

Two-Crib Double-Cantilever Barns

	BUILDER	LOCATION	OVERALL DIMENSIONS		WIDTH		DEPTH				CANTILEVER Primary	
			width	depth	cant	crib	space	cant	crib	#	end	mid
SE32	John Andes	Wears Valley Rd.	50-11	35-10	7-5	12-2	11-9	9-3	17-3	10	6x9	6x11
SE34	Swann	Kyker Ferry Rd.	52-6	36-2	10-2	10-0	12-2	10-1	16-0	4	13x10	13x14
SE36	Smith	Hardin Lane	55-0	32-4	9-6	12-0	12-0	10-1	12-2	10	9x12	9x12
SE37	William Catlett	Old Sevierville Pike	64-0	37-10	10-4	13-0	17-6	9-10	18-2	10	14x12	14x12
SE39	LaFollette	US 411 East	36-4	32-0	1-0	12-0	12-1	10-0	12-0	7	—	—
SE46	———	County Rd. 2486	40-0	30-2	1-3	12-0	11-11	8-1	14-0	15	—	9ø
SE47	———	County Rd. 2486	42-6	30-4	3-1	12-2	12-4	8-1	14-2	14	7x7	7x7
SE48	———	County Rd. 2486	38-9	28-1	2-9	11-9	10-3	8-0	12-1	13	7x7	8x8
SE50	———	US 411	60-0	39-8	9-0	11-6	19-0	9-10	20-0	12	—	—
SE51	James Pickens	DuPont Rd.	61-1	36-0	8-1	13-2	18-7	8-0	20-0	12	7x9	7x18
SE52	Richard Reagan	Reagan Springs Rd.	60-8	36-4	8-6	11-8	20-8	8-0	20-4	12	10x9	12x12
SE53	James Householder	DuPont Rd.	53-3	29-11	9-2	10-10	15-3	7-0	15-8	10	5x8	9x9
SE54	Taylor	Rogers Rd.	70-1	36-2	14-9	12-5	15-9	9-2	11-10	16	12x8	11x14
SE55	———	Dripping Springs Rd.	57-5	34-0	7-10	12-0	17-9	8-1	18-1	12	9x9	9x16
SE56	Rogers	Dripping Springs Rd.	52-8	30-2	8-2	12-0	12-4	8-0	14-2	10	11x7	10x10
SE58	Irving Rogers	Dripping Springs Rd.	52-2	30-5	7-9	11-10	13-0	8-1	14-3	10	9x9	9x13
SE60	James Bohanon	Upper Middle Creek	32-5	25-6	1-5	10-5	9-5	8-1	16-5	6	8x8	8x8
SE62	———	Ridge Rd.	48-6	36-0	5-0	12-10	12-10	8-2	19-10	8	7x7	7x11
SE64	———	Dripping Springs Rd.	61-4	36-2	9-10	11-9	18-1	9-2	17-10	16	11x11	11x13
SE65	———	Dripping Springs Rd.	56-8	30-0	7-0	11-8	19-4	6-6	17-0	14	9x7	9x10
SE66	———	Park Rd.	66-7	29-8	9-1	17-8	17-0	6-0	17-8	11	10x10	9x11
SE67	———	New Era Rd.	48-7	32-0	9-0	10-1	10-5	9-0	14-0	10	5x9	6x13
SE73	Sam T. Proffitt	Rocky Flats Rd.	36-11	31-8	1-3	11-10	12-0	8-10	14-0	12	—	7x9
SE74	John Sharp	Boyd's Creek Rd.	69-10	36-1	8-10	13-10	24-6	8-0	20-1	12	11x12	11x17

DIMENSIONS Secondary		BARN HEIGHTS			L	N	T	F	S	R	C	USE		O	Notes
end	mid	crib	eave	ridge								crib	loft		
9x9	9x12	5-8	6-3	13-4	5	H	Ax	P	H	M	P	St	T	S	JA 1797–1880.
10x10	10x10	6-8	8-6	15-6	6	H	Ax	P	H	M	F	C	H	S	Reversed direction cantilevers.
7x9	9x13	6-5	5-0	10-0	6	H	Ax	P	V	M	G	C	H	NW	
8x9	7x16	6-7	8-4	13-9	5	H	Ax	Q	H	M	F	E	E	SW	WC 1817–95.
8x8	8x14	6-0	5-3	12-6	6	H	Ax	P	H	M	F	St	H	E	
—	10 ø	6-7	6-0	8-6	8	H	Ax	Q	V	M	G	E	T	NE	Asymmetrical. End staircase.
7x6	7x7	6-6	4-9	14-7	6	H	Ax	Q	V	M	F	E	T	SE	Asymmetrical.
—	7ø	6-3	5-2	7-6	6	H	Ax	Q	V	M	F	E	T	NE	Asymmetrical. End staircase.
9x7	9x12	—	5-6	10-6			Ax	Q	H	M	F	P	E	W	Cribs gone.
8x7	8x11	7-9	5-9	14-6	5	H	Ax	Q	V	M	F	St	T	NW	JP b≈1823.
8x8	13x13	5-8	6-9	8-2	5	H	Ax	Q	H	M	G	C	H	SE	RR 1794–1883. 2 pens each crib.
6x9	8x11	7-3	9-0	9-0	5	H	Ax	P	V	M	P	St	T	N	House 1917. JAH 1846–1923.
9x8	9x12	7-1	8-9	14-3	4	H	Ax	P	H	M	F	S	T	N	
9x7	9x12	7-2	6-2	14-0	4	H	Ax	Q	H	M	G	H	E	N	Largest crib log 9x17.
9x6	8x12	6-2	6-6	—	4	H	S	Q	H	M	F	C	T	E	
8x7	10x13	7-5	—	—	5	H	Ax	P	H	M	F	Ho	H	SW	East crib gone.
6x6	6x12	6-0	6-3	8-0	7	H	Ax	Q	H	M	G	St	Ch	S	N. cantilever 1-0. JB 1836–1921.
9x9	9x14	7-2	7-0	12-6	4	H	Ax	Q	H	M	G	C	T	SW	2 pens each crib. Log troughs.
8x9	8x14	5-7	6-0	12-0	4	H	Ax	Q	V	M	P	E	E	S	Extensions 4-3.
9x7	9x7	6-8	6-6	10-6	5	S	Ax	P	V	M	P	C	T	S	
8x9	8x12	7-3	7-6	11-8	6	V	Ax	Q	H	M	F	E	T	W	Dissimilar cribs.
7x7	7x12	6-6	6-3	15-0	4	H	Ax	P	H	M	F	St	H	N	
8x6	9x8	5-6	7-0	14-0	7	H	Ax	P	H	M	F	St	H	S	STP 1857–1914.
10x9	10x16	7-8	8-0	7-0	5	H	Ax	Q	H	M	F	E	E	SE	JS b≈1848.

Table B1 continued
Information Tables for All Barns Surveyed

Two-Crib Double-Cantilever Barns

	BUILDER	LOCATION	OVERALL DIMENSIONS		WIDTH		DEPTH				CANTILEVER Primary	
			width	depth	cant	crib	space	cant	crib	#	end	mid
SE75	Barter	Goose Gap Rd.	60-2	36-0	9-0	12-0	18-2	9-0	18-0	14	9x9	9x11
SE76	———	Goose Gap Rd.	59-6	35-4	9-0	12-0	17-6	8-8	18-0	14	10x7	12x10
SE78	———	Sugar Loaf Rd.	56-4	35-7	9-0	12-0	14-4	9-9	16-1	10	10x8	10x10
SE79	David Cutshaw	Pine Grove Rd.	58-11	36-5	8-9	12-0	18-5	9-1	18-3	14	10x10	10x13
SE81	Caswell T. Ogle	New Era Rd.	45-0	31-8	8-0	9-10	10-2	7-10	16-0	10	8x8	8x10
SE83	Wm. D. Tarwater	Mathis Hollow Rd.	56-10	33-7	9-0	11-11	15-0	8-4	16-11	14	10x9	10x13
SE88	———	Gists Creek Rd.	51-8	31-6	9-2	9-8	14-0	8-11	13-8	12	—	13x14
SE89	Riley J. Ingle	Gists Creek Rd.	55-10	31-9	8-0	12-0	15-10	9-0	13-9	6	6x7	6x13
SE90	———	Gists Creek Rd.	60-5	39-6	10-0	12-0	16-5	9-9	20-0	14	11x6	11x10
SE93	———	Blowing Cove Rd.	60-3	38-0	10-0	12-0	18-3	10-0	18-0	10	8x10	8x14
SE95	———	Old US 411	50-8	40-4	1-0	12-4	24-0	11-2	18-0	11	9x11	9x11
SE102	———	Panther Creek North	50-4	32-0	6-0	10-4	17-8	8-0	16-0	8	8x7	9x13
SE104	———	Union Valley Rd.	55-1	34-10	8-2	10-2	16-5	9-2	16-6	6	—	10x8
SE105	Cutshaw	Long Branch Rd.	59-1	30-0	7-6	9-11	24-3	8-0	14-0	12	11x5	11x13
SE106	———	Knob Creek Rd.	56-1	29-8	8-0	11-10	16-5	6-6	16-8	10	8x7	8x12
SE107	Johnson	Low Posse Hollow	64-8	34-10	8-1	13-9	21-0	7-4	20-2	12	11x11	12x10
SE110	———	Flat Creek Rd.	39-10	32-0	2-0	11-10	12-2	9-9	12-6	10	10x10	10x10
SE112	———	Long Branch Rd.	59-8	36-0	—	13-0	33-8	10-0	16-0	12	9x12	11x15
SE113	Ingle	Ingle Hollow Rd.	51-4	33-6	7-4	10-10	15-0	9-9	18-0	12	10x9	10x9
SE114	Ingle	Ingle Hollow Rd.	53-10	31-9	6-7	12-3	16-2	9-0	13-9	12	8x6	8x10
SE115	———	Mathis Hollow Rd.	35-8	31-10	1-9	10-0	12-2	8-11	12-0	8	7x9	7x9
SE116	———	Allensville Rd.	40-7	29-8	5-5	12-0	5-9	7-10	14-0	9	8x5	7x6
SE117	———	Robertson Rd.	53-0	30-5	9-0	10-10	13-4	9-4	11-9	10	8x8	8x8
SE118	———	Robertson Rd.	55-10	38-2	10-0	11-9	12-4	10-0	18-2	12	8x8	8x13

DIMENSIONS Secondary end	mid	BARN HEIGHTS crib	eave	ridge	L	N	T	F	S	R	C	USE crib	loft	O	Notes
9x8	9x12	6-8	10-0	12-6	5	H	Ax	Q	H	M	F	C	T	S	Extensions 3-1.
9x7	7x11	6-3	9-2	10-6	5	H	Ax	Q	H	M	F	H	H	NE	Extensions 3-3.
13x8	12x10	7-6	4-5	9-0	6	H	Ax	P	H	M	F	St	St	W	
7x9	7x11	6-6	6-0	11-0	5	H	Ax	—	H	M	F	C	H	NW	Extensions 3-3. DC 1849–1936.
7x7	9x10	7-0	7-4	12-6	6	H	Az	Q	H	M	F	St	HT	S	Built 1890. CTO b≈1843.
8x8	8x11	6-4	8-11	9-0	5	H	Ax	P	H	M	G	St	T	SW	Extensions 2-6. WDT 1855–1924.
9x6	9x11	6-4	7-6	11-0	5	H	Ax	Q	H	M	G	Ho	H	S	Extensions 3-6.
7x6	9x10	6-9	4-6	9-6	5	H	Ax	Q	H	M	F	St	HT	W	Extensions 3-0. RJI 1875–1966.
8x8	8x11	6-0	7-1	14-1	5	H	Ax	Q	H	M	F	C	H	N	Extensions 3-0.
7x9	8x16	7-0	11-0	20-0	4	H	Ax	P	H	M	G	A	HT	NE	Continuous scarf-joint sill.
6x5	6x14	7-3	6-6	7-6	5	H	Ax	P	V	M	F	HT	HT	S	
7x4	7x9	5-2	6-1	9-0	5	H	Ax	Q	V	M	F	tu	H	SW	Extensions 2-2.
8x9	8x14	6-0	5-9	8-0	5	H	Ax	Q	H	M	F	St	HT	NW	
10x4	10x11	5-2	6-4	10-0	5	H	Ax	Q	H	M	F	A	HT	N	Extensions 4-0.
9x4	9x10	5-8	7-5	16-0	5	H	Ax	Q	H	M	P	HT	HT	S	Extensions 3-0.
12x12	12x11	7-2	5-2	11-0	4	H	Ax	P	H	M	F	H	E	NW	Skid notch on outer end.
7x6	7x13	6-4	6-8	14-6	5	H	Ax	Q	H	M	G	T	T	SW	
9x10	9x14	7-6	7-9	10-0	4	H	Ax	Q	H	M	F	St	HT	S	Inverted double cantilever barn.
9x5	10x12	6-0	6-9	9-0	5	H	Ax	Q	H	M	P	St	E	NW	Extensions 3-0.
8x7	8x10	6-8	5-6	13-0	5	H	Ax	P	H	M	G	Ho	H	SW	Extensions 3-3.
8x9	8x11	6-0	6-7	11-0	5	H	Ax	P	H	W	P	St	T	SW	
6x6	6x12	5-9	5-10	11-8	7	V	Ax	Q	H	M	F	E	E	S	Continuous scarf-joint primary.
7x7	7x9	6-8	6-4	15-8	6	H	Ax	P	H	M	F	St	T	SE	Continuous scarf-joint primary.
8x7	8x13	6-3	8-5	9-0	5	H	Ax	Q	V	M	F	E	H	NE	SE crib has 8 secondaries.

Table B1 continued
Information Tables for All Barns Surveyed

Two-Crib Single-Cantilever Barns

	BUILDER	LOCATION	OVERALL DIMENSIONS		WIDTH		DEPTH				CANTILEVER Primary	
			width	depth	cant	crib	space	cant	crib	#	end	mid
SE122	Samuel P. Flynn	Kellum Creek Rd.	50-8	36-0	8-10	12-0	9-0	9-0	18-0	12	8x9	8x15
SE124	———	———	60-6	35-6	8-0	12-0	20-6	7-6	20-6	10	8x9	9x14
SE133	John Conatsher	LaFollette Rd.	36-10	32-0	1-6	12-0	9-10	10-0	12-0	8	7x12	7x12
SE136	Horatio Butler	Mount Zion Rd.	48-2	31-11	7-0	11-5	11-4	9-2	13-9	14	8x8	9x12
SE138	Robert Henderson	Roberts Schoolhouse	36-0	30-2	1-0	12-0	10-0	9-1	12-0	8	7x11	7x12
SE139	William Sutton	Roberts Schoolhouse	49-7	34-2	9-9	10-2	9-9	9-0	16-2	10	—	—
SE146	———	Thomas Cross Rd.	48-4	37-3	3-0	11-7	19-2	9-10	17-7	9	11x8	11x8
SE150	Jacob H. Seaton	Necum Hollow Rd.	33-10	30-6	2-1	10-0	9-8	9-3	12-0	9	7x8	7x13
SE152	Jacob H. Seaton	Necum Hollow Rd.	53-2	32-8	9-3	12-0	10-8	9-4	14-0	10	9x8	9x12
SE156		Shinbone Rd.	49-10	35-10	8-0	11-11	10-0	9-0	17-10	10	9x12	9x12
SE160	———	Pine Grove Rd.	52-3	41-10	3-5	10-9	13-11	9-0	13-10	8	8x8	8x13
SE161	Jenkins	Sugar Loaf Rd.	48-0	30-0	8-0	10-0	12-0	9-0	12-0	10	8x6	8x9
SE162	Gibson	Sugar Loaf Rd.	54-1	34-8	8-4	12-0	13-8	8-4	18-0	12	11x9	11x13
SE163	———	Sugar Loaf Rd.	37-4	24-0	4-0	9-8	10-0	6-0	12-0	10	8x7	8x9
SE168	James R. Tarwater	Mathis Hollow Rd.	54-4	32-0	8-5	11-9	14-0	8-0	16-0	12	11x8	11x12
SE172	John Ogle	Roberts Schoolhouse	52-1	32-0	8-5	12-0	11-3	9-0	14-0	10	8x12	8x12
SE173	W. Philip Roberts	Seaton Springs Rd.	51-7	34-8	8-9	11-10	10-5	8-9	17-2	12	12x9	13x11
SE178	G. Decator Snapp	Tiger Valley Rd.	57-10	36-8	7-9	14-0	14-4	9-4	18-0	9	9x8	9x8
SE179	———	Reagon Town Rd.	47-3	32-0	8-9	10-0	10-0	9-0	14-0	11	7x7	7x13
SE180	James H. Butler	McMahan Rd.	48-6	31-10	7-0	11-6	11-6	9-0	13-10	10	8x7	8x9
SE183	———	[Appalachia Museum]	62-10	41-5	10-6	12-8	16-6	10-4	20-9	10	10x9	14x10

| DIMENSIONS | | BARN HEIGHTS | | | | | | | | | | | USE | | | |
| Secondary | | | | | | | | | | | | | | | | |
end	mid	crib	eave	ridge	L	N	T	F	S	R	C	crib	loft	O	Notes
7x8	9x13	6-8	10-0	12-6	5	H	Ax	P	H	M	P	E	E	N	Loft not original. SPF 1852–1930.
8x10	8x10	6-7	6-8	15-6	5	H	Ax	Q	H	M	F	C	H	NE	
8x8	8x13	5-6	6-0	16-8	7	H	Ax	P	H	M	F	C	H	NW	
10x8	10x13	7-2	6-2	14-6	6	H	Ax	Q	H	M	P	St	T	N	Has log house. HB 1797–1849.
7x9	8x12	6-0	6-8	10-3	6	H	Ax	Q	H	M	F	St	St	S	RH 1841–1923.
6x5	6x10	8-0	7-6	12-6	—	—	Ax	Q	H	M	G	St	HT	NW	Cribs gone. WS 1866–1941.
9x7	9x9	7-0	7-0	19-9	6	H	Ax	Q	H	M	G	C	H	NE	Continuous scarf-joint primary.
7x7	7x11	7-7	6-8	10-6	6	H	Ax	Q	H	M	G	St	HT	SE	Floor over aisle. JHS 1856–1909.
9x9	9x12	6-7	9-2	12-6	5	H	Ax	Q	H	M	F	E	E	NE	JHS 1856–1909.
6x6	6x10	6-6	9-3	12-6	5	H	Ax	Q	H	M	G	St	HT	NE	"H.Tedford 1888" on crib log.
7x6	7x10	6-10	6-3	15-6	5	H	Ax	Q	H	M	F	St	T	W	
9x6	9x10	5-8	6-0	6-0	5	H	Ax	P	H	M	G	E	E	W	
8x9	8x12	6-6	6-0	11-8	4	H	Ax	P	H	M	P	E	E	E	
7x4	7x8	5-8	5-0	10-6	6	H	Ax	P	H	M	G	C	H	N	
11x6	11x12	6-4	6-5	11-8	5	H	Ax	P	H	M	F	Ho	HT	N	Extensions 2-2. JRT 1851–1932.
8x6	7x10	6-3	9-0	12-6	6	H	Ax	Q	H	M	F	St	E	NW	JO 1848–1930.
7x8	8x9	6-4	7-9	14-0	5	H	Ax	Q	H	M	F	E	E	W	Side-door cribs. WPR 1803–75.
8x7	8x12	8-0	6-9	8-0	—	—	Ax	P	H	M	G	St	H	SE	Cribs replaced by concrete block.
6x6	7x9	6-6	6-6	12-6	5	H	Ax	Q	H	M	F	C	H	E	
8x8	8x12	5-8	7-0	11-3	5	H	Ax	Q	H	M	G	C	St	S	Skid notches. JHB 1849–1930.
9x10	9x13	6-8	6-0	15-0	5	H	Ax	Q	V	W	G	St	H	?	From Boyd's Creek. 3 primaries.

Table B1 continued
Information Tables for All Barns Surveyed

Two-Crib Single-Cantilever Barns

	BUILDER	LOCATION	OVERALL DIMENSION		WIDTH		DEPTH		Secondary	CANTILEVER DIMENSIONS	
			width	depth	crib	space	cant	crib	#	end	mid
BL4	Kidd	Nails Creek Rd.	48-2	34-4	14-0	20-2	7-0	20-4	12	7x8	7x13
BL21	John A. Graves	Chilhowee Rd.	36-3	20-0	11-10	12-7	4-0	12-0	8	6x11	11x10
BL31	——	Sheets Mill Rd.	31-0	31-0	11-0	10-0	10-0	11-0	4	6x5	6x8
BL32	William P. Brown	Miser Station Rd.	38-8	30-10	11-11	15-10	7-5	16-0	4	7x9	7x12
BL40	Owens	Honeysuckle Rd.	36-0	31-0	12-0	12-0	7-6	16-0	4	7x7	7x13
BL41	Bartley Langford	Ridge Rd.	35-9	27-6	12-0	11-9	8-9	10-0	4	6x6	6x8
BL42	W. Elmer Wright	Jennings Rd.	39-6	31-6	11-10	15-10	7-9	16-0	4	7x7	6x14
BL43	Jimmy Hall	Middle Settlements	38-4	36-0	14-0	10-4	9-0	8-0	4	8x7	15x8
BL44	Pink Stoner	Day Rd.	31-6	30-6	9-5	12-8	10-0	10-6	4	7ø	7ø
BL48	Billy Kear	U.S. 411	31-4	30-1	10-8	10-10	8-9	12-7	4	6x9	6x11
BL52	Edgar Roy Walker	Bethlehem Rd.	32-1	18-4	10-8	10-9	2-11	12-4	6	8x6	8x9
BL54	Dan Henry	Ellejoy Rd.	40-10	30-0	12-3	16-4	8-0	14-0	8	9x9	9x9
BL55	Philander Clemens	Melonie Rd.	30-1	26-2	10-0	8-5	8-0	10-2	—	6x6	6x8
BL60	Robert H. Wright	Cold Springs Rd.	34-3	26-0	12-0	10-3	8-0	10-0	8	8ø	8ø
BL61	Anderson Keeble	Cold Springs Rd.	33-6	25-10	11-11	9-8	8-0	9-10	8	6x6	6x6
BL63	James W. Burns	Loop Rd.	39-0	26-9	11-10	15-4	6-5	13-11	8	6x6	6x13
BL64	Frank Gridicks	E. Miller's Cove Rd.	18-10	28-0	18-10	—	8-0	12-0	5	7x7	7x10
BL67	Keeler	Tommy Pack Rd.	35-9	28-0	12-0	11-9	8-0	12-0	6	7x7	8x12
BL68	Elie 'Aley' Caylor	Frogtown Rd.	30-2	24-0	10-0	10-2	7-0	10-0	6	7x8	7x8
BL69	Henry Tipton	Owenby Rd.	24-9	26-10	24-9	—	7-4	12-2	5	7x7	8x10
BL70	Art Emert	U.S. 321	35-8	28-3	11-10	12-0	7-11	12-5	8	8x6	9x12
BL73	Susie Henry	White's Mill Rd.	24-3	24-0	6-0	8-0	6-0	12-0	4	6x6	6x6
BL83	McDonald	Ridge Rd.	34-1	27-2	12-0	10-1	6-7	14-0	4	7x5	7x10
BL92	Silas McMillan	Regan Mountain Rd.	28-4	28-0	9-8	9-0	8-0	12-0	6	8x6	8x11

| BARN HEIGHTS | | | | | | | | | | | USE | | | |
crib	eave	ridge	L	N	T	F	S	R	C	crib	loft	O	NOTES
6-9	6-0	10-6	6	V	Ax	P	V	M	F	C	HT	—	c. 1900
5-8	5-0	5-0	5	V	Ax	P	V	M	F	C	E	SW	Extensions 0-10. c. 1910. JAG 1875–1945.
6-2	4-0	13-0	7	S	Ax	P	H	M	P	E	E	SE	
7-5	6-0	11-0	6	H	Ax	P	V	M	P	E	E	NE	1901–2. WPB 1869–1935.
6-0	6-8	13-0	5	H	Ax	Q	H	M	F	C	HT	SE	Pre-1900.
5-9	6-0	9-6	6	H	Ax	Q	V	M	F	St	H	SE	c. 1885–95.
7-0	—	—	5	H	Ax	—	—	—	P	E	E	—	Loft gone. Pre-1885. WEW 1834–97.
6-4	6-5	9-0	4	H	Ax	Q	V	M	P	St	T	NW	Pre-1895.
5-6	—	—	6	S	Saw	P	V	M	F	St	St	—	Ridge parallels central space. 1932.
6-2	7-2	15-6	7	H	Ax	Q	H	M	F	Ho	H	NW	Ridge parallels central space. PS 1897–1954.
5-5	6-4	4-6	5	V	Ax	—	V	M	F	C	E	W	c. 1901. ERW 1883–1972.
7-0	12-0	4-0	7	V	Ax	Q	V	M	F	E	E	NW	c. 1890.
5-0	5-4	5-6	6	V	Ax	Q	V	M	P	C	E	SE	Extensions 1-1. Pre-1910.
6-9	6-8	8-0	7	V	Ax	P	V	M	F	St	St	S	Cribs open to center space. RHW 1906–77.
5-9	5-6	11-0	7	V	Ax	Q	H	M	F	C	T	SE	Pre-1920.
5-3	5-3	10-6	5	H	Saw	Q	V	M	P	E	E	W	Exts. 4-0. Cribs hewn. JWB 1849–1920.
6-5	5-7	11-2	6	H	Ax	Q	V	M	G	E	T	W	Single crib with 2 pens. Pre-1910.
6-0	6-2	11-3	6	S	Ax	Q	H	M	F	C	H	E	Pre-1920.
5-4	8-0	5-0	6	S	Ax	P	V	M	F	H	H	S	West crib gone. Pre-1900. EC 1809–94.
6-0	5-3	—	7	S	Ax	Q	H	M	F	E	E	N	Single crib with 2 pens. c. 1902.
5-7	9-10	4-0	5	S	Ax	Q	V	M	F	St	H	NW	c. 1890.
5-0	3-3	10-6	5	HV	Ax	Q	V	M	P	E	E	SE	Unequal cribs. c. 1901.
6-8	—	—	5	S	Ax	—	V	M	G	St	H	—	Rebuilt after wind damage to loft. c. 1885.
6-9	5-0	12-6	6	H	Ax	Q	H	M	F	St	H	W	c. 1892. SMC 1844–1930.

Table B1 continued
Information Tables for All Barns Surveyed

Two-Crib Single-Cantilever Barns

	BUILDER	LOCATION	OVERALL DIMENSION		WIDTH		DEPTH		Secondary #	CANTILEVER DIMENSIONS	
			width	depth	crib	space	cant	crib		end	mid
BL94	Otis T. Brewer	Bingham Rd.	44-11	32-1	15-9	15-2	8-2	15-9	4	10x6	10x6
BL96	Crye	Brick Mill Rd.	17-7	26-10	17-7	—	8-2	10-6	2	7x8	7x8
BL97	Bert A. Carpenter	Bartshed Rd.	36-10	30-0	10-9	15-7	8-0	14-0	4	7x11	7x10
BL98	Mack Cagley	Six Mile Rd.	26-0	20-0	12-0	5-5	5-0	15-0	10	8x9	—
BL103	James Anderson	Bob Irwin Rd.	35-0	27-8	11-6	12-0	8-0	11-8	4	9x7	9x11
BL104	George R. Gardner	Mutton Hollow Rd.	39-9	35-10	12-0	15-9	9-0	11-10	24	9x6	9x6
CO3	———	Indian Creek Rd.	34-2	29-6	11-3	11-8	8-2	13-2	4	9x7	9x9
CO4	———	TN 32 at Cosby	35-6	28-0	11-9	12-0	8-3	11-6	10	8x7	—
CO5	———	TN 32 at Cosby	37-0	27-0	11-10	13-4	7-6	12-0	4	9ø	9ø
CO6	———	Happy Hollow Rd.	40-0	32-4	14-0	12-0	9-2	14-0	4	8x7	8x10
JE1	———	Service Rd. off I-40	36-7	33-6	12-0	12-6	9-7	14-4	4	8x9	7x10
JO3	Slimp	TN 167 at Butler	38-6	31-0	12-1	14-4	9-2	12-3	6	—	8x9
JO5	———	TN 167 at Butler	35-7	30-7	11-5	12-9	8-5	13-9	4	—	7x9
JO6	Thomas Johnson	TN 91	36-6	40-10	12-0	12-6	11-5	18-0	4	—	—
KN5	Callie L. Bales	Dave Smith Rd.	49-7	39-5	14-10	20-0	9-9	19-11	4	12x11	12x18
SE4	Sanford Allen	Middle Creek Rd.	48-5	31-11	19-8	9-3	10-0	12-1	12	7x7	7x7
SE21	John A. Seaton	Bullman Rd.	34-3	32-3	12-0	10-3	9-0	14-3	10	7x8	7x14
SE27	John Butler Jr.	Richardson Cove Rd	38-2	31-1	14-1	10-0	8-6	14-1	10	8x8	7x11
SE30	Earle Headricks	Line Springs Rd.	38-2	29-6	12-0	14-2	8-0	13-6	—	9x9	9x12
SE31	———	Walden Creek Rd.	30-11	24-8	10-3	10-3	6-4	12-0	6	8x7	7x8
SE35	Smith	Hardin Lane	41-11	31-8	14-10	12-3	9-10	12-0	10	6x8	6x8

| BARN HEIGHTS | | | | | | | | | | USE | | | |
crib	eave	ridge	L	N	T	F	S	R	C	crib	loft	O	NOTES
6-2	5-3	10-6	6	V	Ax	Q	H	M	P	T	T	—	Pre-1900. OTB 1898–1971.
5-0	6-8	—	7	S	Ax	Q	V	M	P	St	St	SE	c. 1900.
7-4	6-6	6-6	6	V	Ax	Q	V	M	G	C	T	—	Continuous primary and sill. BAC1889–1968.
6-4	10-6	8-0	5	H	Ax	Q	HV	M	F	St	H	E	Reused logs. No rear cantilever. 1870s.
6-0	5-7	8-9	7	H	Ax	Q	H	M	P	E	T	SW	Cribs open to center. JA 1804–85.
6-6	6-3	9-0	5	H	Ax	Q	H	M	P	C	H	SE	Rev. prim. w/ 12 second. GRG 1852–1940.
6-5	4-2	12-6	6	H	Ax	P	H	M	G	C	T	SE	
6-0	3-6	4-0	7	V	Ax	P	H	M	P	St	H	NE	Log trough in NW crib.
—	—	—	7	V	Ax	P	V	M	F	St	—	SW	
6-9	4-6	13-6	8	V	Ax	P	V	M	G	St	H	E	
5-6	7-6	12-6	7	V	Ax	Q	H	W	P	E	E	NE	Circular-sawed loft frame.
6-6	5-9	10-0	5	S	Ax	Q	V	M	F	St	T	W	
6-5	4-7	6-0	—	V	Ax	—	V	M	F	C	T	SE	Two story cribs.
7-0	5-3	15-2	8	V	Ax	—	V	M	F	T	T	E	Two story cribs.
5-8	5-0	8-9	4	H	Ax	Q	H	M	G	C	HT	N	Said to be built c. 1865.
6-6	6-10	11-8	6	H	Ax	P	H	M	P	St	St	N	Moved from Douglas 1942. SA 1848–1934.
7-8	6-3	12-6	5	H	Ax	P	H	M	F	CH	T	E	1-0 cantilevers on south. JAS 1838–1910.
6-4	7-8	12-6	8	H	Ax	P	H	M	F	St	H	SE	
6-4	7-6	10-6	6	H	Ax	Q	V	M	F	E	E	SW	
6x5	5x0	8x0	6	S	Ax	P	H	M	F	St	E	N	Extensions 1-9.
6-0	6-3	12-6	7	H	Ax	P	H	M	G	St	H	NE	

Table B1 continued
Information Tables for All Barns Surveyed

Two-Crib Single-Cantilever Barns

	BUILDER	LOCATION	OVERALL DIMENSION		WIDTH		DEPTH		Secondary	CANTILEVER DIMENSIONS	
			width	depth	crib	space	cant	crib	#	end	mid
SE41	——	U.S. 411	31-0	33-2	11-10	7-4	9-7	14-0	8	8x6	8x12
SE42	Whaley	U.S. 411	35-6	31-8	11-9	12-0	9-10	12-0	8	6x7	6x10
SE43	John Shultz	Rocky Flats Rd.	33-2	30-2	11-2	10-10	9-1	12-1	—	—	9x7
SE44	Webb(?)	Henrytown Rd.	25-10	28-4	10-1	5-5	7-2	14-0	9	8x6	9x8
SE45	George G. Allen	Jones Cove Rd.	36-1	29-8	11-7	12-4	7-10	14-0	9	8x6	8x12
SE59	——	Upper Middle Creek	34-0	30-2	12-0	10-0	9-1	12-0	8	7x9	7x12
SE61	John Marshall	Upper Middle Creek	35-10	30-2	11-10	12-2	9-0	12-2	8	7x8	7x13
SE63	——	U.S. 411	34-7	25-11	11-8	11-3	7-6	11-11	4	9x7	9x12
SE68	——	U.S. 321/Wears Vall.	33-11	30-0	12-0	9-11	8-0	12-0	4	8x7	8x11
SE69	King	U.S. 321/Wears Vall.	30-10	28-8	10-5	10-0	8-2	12-4	7	8x7	7x8
SE70	George A. King	U.S. 321/Wears Vall.	40-0	29-6	14-0	12-0	7-10	13-10	7	10x7	8x11
SE72	John Williams	Rocky Flats Rd.	31-8	27-11	9-10	12-0	7-0	13-11	4	8x8	8x4
SE77	Henry Tipton	Sugar Loaf Rd.	36-2	40-0	12-0	12-2	10-0	20-0	—	—	—
SE80	——	New Era Rd.	30-2	23-3	9-6	11-2	7-0	9-3	6	5x8	7x9
SE82	——	New Era Rd.	39-8	30-0	13-9	12-2	9-0	12-0	6	8x6	9x14
SE84	——	Pine Grove Rd.	33-4	25-10	11-0	11-4	8-0	9-10	6	7x11	7x11
SE85	Levi Tarwater	Pine Grove Rd.	33-11	28-6	10-2	13-10	8-0	12-6	6	8x6	9x10
SE86	——	Panther Creek Rd.	41-0	31-11	12-6	16-0	8-10	14-3	8	9x7	9x10
SE87	——	Panther Creek Rd.	35-8	30-2	9-10	16-0	8-2	14-0	8	8x5	8x9
SE91	Rule	Rule Rd.	40-6	32-2	12-0	16-6	9-2	13-10	8	9x6	9x8
SE92	Delozier	Gists Creek Rd.	31-3	32-3	8-6	14-3	8-10	14-7	4	5x8	7x8
SE94	——	U.S. 411	34-1	30-0	9-8	14-9	9-0	12-0	9	9x7	8x8
SE97	Henry G. Hodges	McCleary Rd.	40-0	32-4	11-9	16-0	9-0	14-4	—	—	10x9
SE98	——	McCleary Rd.	50-8	32-4	16-0	18-8	7-1	18-3	4	9x12	9x16

| BARN HEIGHTS | | | | | | | | | | USE | | | |
crib	eave	ridge	L	N	T	F	S	R	C	crib	loft	O	NOTES
6-4	5-5	9-7	6	H	Ax	P	H	M	G	P	E	SE	
6-0	5-0	10-6	7	H	Ax	P	H	M	F	St	T	NW	
6-8	5-0	16-0	6	H	Ax	P	V	M	G	C	T	SE	Cribs open to center. c. 1914. Sawed cribs.
6-0	5-3	12-6	6	V	Ax	P	H	M	P	St	H	E	
7-2	7-0	10-0	6	V	Ax	P	H	M	G	St	HT	N	GGA 1888–1957.
7-3	6-6	11-3	6	H	Ax	P	H	M	F	St	T	NW	
6-9	7-6	12-6	7	H	Ax	Q	H	M	G	C	H	W	JM 1829–92.
5-6	5-10	7-0	5	V	Ax	P	H	M	G	C	HT	S	
6-6	7-0	—	6	H	Ax	Q	H	M	F	C	H	N	Roof raised above original purlins.
6-6	6-8	9-2	5	H	Ax	P	H	M	F	C	H	S	Adze marks also visible.
6-6	6-8	8-4	5	H	Ax	Q	H	M	F	Ho	H	S	GAK 1806–57.
7-0	—	—	5	H	Ax	—	V	M	G	C	H	NE	Moved 100 yds. south in 1948. JW b. 1843.
6-0	—	—	5	H	Saw	Q	H	M	P	C	H	SE	Cantilevers gone. HT 1897–1952.
6-0	—	—	5	H	Ax	P	H	M	P	E	E	SW	Loft rebuilt with utility poles.
6-4	6-0	15-0	6	H	Ax	Q	H	W	P	C	T	S	Cribs open to center space.
6-0	5-6	9-6	7	V	Ax	Q	H	M	F	E	HT	N	Cribs open to center. "1922" painted on crib.
6-6	8-6	6-0	6	H	Ax	P	V	M	F	A	T	NW	Cribs open to center. LT 1898–1959.
4-9	5-5	11-0	5	H	Ax	P	H	M	F	C	E	N	Extensions 2-11.
7-4	4-6	11-6	7	V	Ax	P	V	M	F	C	HT	S	Extensions 2-5.
6-6	4-6	16-3	5	H	Ax	Q	V	M	F	St	E	NW	Extensions 3-0.
6-4	5-4	9-0	8	H	Ax	Q	V	M	F	C	H	S	
6-9	5-4	9-0	5	H	Ax	—	V	M	DF	H	H	SW	
7-7	5-2	9-6	6	S	Ax	Q	H	M	G	T	T	SE	HGH 1879–1955.
7-0	—	—	5	V	Ax	—	H	M	P	H	H	N	West crib 2 story log. East crib gone.

Table B1 continued
Information Tables for All Barns Surveyed

Two-Crib Single-Cantilever Barns

		OVERALL DIMENSION		WIDTH		DEPTH		Secondary	CANTILEVER DIMENSIONS	
BUILDER	LOCATION	width	depth	crib	space	cant	crib	#	end	mid
SE99 ———	McCleary Rd.	35-9	32-3	11-2	13-4	8-0	16-3	6	6x12	9x12
SE100 ———	Bob Hollow Rd.	35-2	30-4	11-3	12-8	8-8	13-0	6	6x6	10x6
SE101 ———	Panther Creek North	32-8	25-10	10-0	12-8	7-10	10-2	8	12x8	12x8
SE103 ———	Union Valley Rd.	34-2	32-4	12-6	9-2	8-5	15-6	4	11x9	11x12
SE108 ———	Low Posse Hollow	34-0	28-6	12-0	10-0	7-3	14-0	4	7x7	7x10
SE109 ———	Hetty Creek Rd.	35-10	32-3	10-0	10-0	9-3	14-0	9	8x8	7x17
SE111 ———	Flat Creek Rd.	33-10	30-0	11-7	10-3	9-0	12-0	9	8x7	8x13
SE119 Will C. Jones	Powder Springs Rd.	37-4	28-1	9-3	18-5	5-11	16-3	4	6x6	7x8
SE120 ———	Allensville Rd.	40-7	28-4	19-10	10-9	9-4	9-8	9	7x8	7x9
SE121 Cap Wynn	Old TN 66	30-2	26-6	10-0	10-2	8-0	10-6	4	8x6	10x13
SE123 ———	Kellum Creek Rd.	35-0	28-4	12-0	11-0	8-2	12-0	6	11x6	10x8
SE125 Bryan	Bryan Rd.	48-8	29-10	19-8	9-4	8-9	11-9	[8]	6x10	6x11
SE126 ———	Bryan Rd.	38-10	34-10	12-4	14-2	9-6	15-10	4	7x7	7x11
SE128 ———	Lane Hollow Rd.	34-8	32-0	12-0	10-8	10-0	12-0	10	7x9	7x11
SE130 Fox	Happy Hollow Rd.	35-10	28-10	11-10	12-2	8-5	12-0	6	7x8	7x14
SE131 Robert Henderson	Happy Hollow Rd.	35-10	30-4	11-8	12-6	9-4	11-8	6	8x10	7x17
SE132 ———	Lafollette Rd.	43-9	32-10	16-9	10-3	10-0	12-0	10	10x7	10x15
SE134 Decator Conatsher	Lafoillette Rd.	32-10	32-9	11-8	9-6	10-3	12-3	2	8x9	9x13
SE135 John W. Whitehead	Lafollette Rd.	34-2	30-2	12-0	10-2	9-6	10-10	8	6x7	7x6
SE137 Robert Marshall	Middle Creek Rd.	35-0	30-0	11-10	10-0	8-0	14-0	8	8x8	8x11
SE141 Gilbert Thompson	Maple Branch Rd.	31-8	28-0	9-7	12-7	8-1	11-0	6	8x6	9x11
SE143 ———	Thomas Rd.	35-9	32-0	12-0	11-9	9-0	14-0	10	7x8	7x12
SE144 ———	Thomas Cross Rd.	29-11	27-9	11-8	6-7	8-0	9-9	9	7ø	7ø
SE145 ———	Fox Cemetery Rd.	40-9	32-4	12-0	16-9	10-2	12-0	11	7ø	7ø

| BARN HEIGHTS | | | L | N | T | F | S | R | C | USE | | O | NOTES |
crib	eave	ridge								crib	loft		
6-8	5-3	—	6	HV	Ax	Q	H	M	F	H	HT	NE	NW crib is older than SE crib.
6-7	—	—	7	V	Ax	P	V	M	G	R	R	S	
5-7	6-0	12-0	6	H	Ax	Q	V	M	F	Ho	H	SW	Extensions 1-8.
8-0	5-8	10-0	5	H	Ax	P	V	M	—	C	T	NW	
5-9	8-0	8-0	6	H	Ax	Q	V	M	F	St	St	E	
5-9	6-4	12-6	5	H	Ax	Q	H	M	G	H	H	S	East crib 15-0 wide.
7-0	6-3	12-6	7	V	Ax	?	H	M	G	St	H	SW	
6-8	4-4	6-10	8	V	Ax	Q	H	M	F	St	H	S	House built c. 1860. WCJ 1842–1904.
6-6	5-0	10-0	5	H	Ax	P	V	M	F	St	H	SW	NW crib has 2 pens and 6 cantilevers.
5-0	5-9	7-0	7	V	Ax	P	H	M	P	St	E	NE	
5-6	7-0	7-0	6	H	Ax	Q	H	M	P	C	H	S	
6-0	6-8	6-8	6	H	Ax	Q	V	M	P	C	H	NE	SE crib gone.
5-6	4-9	11-0	5	H	Ax	Q	V	M	F	St	T	E	Continuous sill between cribs.
5-0	5-8	12-6	4	H	Ax	Q	H	M	P	E	E	SW	
4-6	7-3	12-6	4	H	Ax	Q	H	M	P	E	H	—	
6-5	5-9	7-6	5	H	Ax	Q	H	M	F	C	H	NW	Continuous primary beam. RH 1841–1923.
6-0	7-6	13-0	5	H	Ax	Q	H	M	F	C	T	NW	
6-0	7-10	12-6	7	H	Ax	Q	H	M	P	C	HT	S	
6-0	7-0	12-6	7	H	Ax	Q	H	M	F	E	E	N	Adjacent house c. 1895.
7-2	7-0	7-6	6	H	Ax	Q	H	M	G	Ho	H	NE	RM 1830–88.
5-9	6-8	8-0	6	H	Ax	P	V	M	G	A	HT	NE	GT 1879–1952.
5-8	5-2	15-0	6	H	Ax	Q	H	M	F	E	E	E	Continuous primary beam. Heavily repaired.
5-1	6-0	8-0	6	HV	Ax	P	HV	M	F	St	T	—	Continuous primary. Crude construction.
7-2	6-8	7-6	7	H	Ax	—	V	M	F	A	HT	—	Continuous primary beam and sill.

Table B1 continued
Information Tables for All Barns Surveyed

Two-Crib Single-Cantilever Barns

		OVERALL DIMENSION		WIDTH		DEPTH		Secondary	CANTILEVER DIMENSIONS	
BUILDER	LOCATION	width	depth	crib	space	cant	crib	#	end	mid
SE147 ———	Thomas Cross Rd.	26-2	22-1	10-6	5-0	7-0	8-1	6	7ø	7ø
SE148 ———	King Branch Rd.	35-1	28-5	12-10	9-4	7-10	12-9	7	7x6	7x12
SE151 Jesse Atchley	Necum Hollow Rd.	30-5	29-11	9-11	10-7	4-11	10-1	4	6x8	6x12
SE153 ———	Buckhorn Rd.	39-8	29-10	11-10	16-0	8-0	13-10	11	7x6	7x10
SE154 ———	Campbell Branch Rd.	34-4	37-5	12-2	10-2	8-0	11-5	6	7x7	7x10
SE155 Sneed?	King Hollow Rd.	36-3	32-3	11-10	12-7	10-4	11-7	4	9ø	9ø
SE157 ———	Cummings Chapel Rd	36-0	30-0	12-0	12-0	9-0	12-0	10	5x8	5x16
SE158 Bogart	Cummings Chapel	40-0	30-0	14-0	12-0	9-0	12-0	8	5x6	5x14
SE159 ———	Pearl Valley Rd.	39-7	32-0	13-10	11-11	8-0	16-0	10	7x5	7x6
SE167 Robert Henderson	Witt Hollow Rd.	51-1	33-6	19-10	11-5	9-10	13-10	11	7x8	8x12
SE169 ———	Valley Rd.	36-0	29-9	12-0	12-0	8-0	13-9	6	8x9	8x8
SE170 Arthur Suttles	Valley Rd.	35-6	27-10	12-0	11-6	8-0	11-10	8	7x7	7x7
SE174 ———	Bomertown Rd.	39-6	24-0	11-2	7-2	6-8	8-8	7	9x7	9x9
SE175 ———	Shell Mountain Rd.	31-6	28-6	14-0	9-8	6-4	15-10	15	7x8	7x8
SE176 ———	Flat Creek Rd.	39-2	31-2	18-0	10-10	10-0	10-6	10	8x7	8x12
SE181 ———	Millwood Drive	34-11	23-10	12-3	10-5	5-10	12-2	6	6x7	6x7
SE182 Bruce Stinett?	Ridge Rd.	26-10	25-1	8-9	9-4	6-3	12-7	6	6x6	6x10
UN3 George Bowman	U.S. 19-23	46-8	39-8	15-4	16-0	9-10	20-0	4	—	11x10
WA3 Garst	Old TN 34-U.S. 11E	60-6	39-8	18-0	24-6	9-9	20-2	6	—	12x15

BARN HEIGHTS											USE			
crib	eave	ridge	L	N	T	F	S	R	C	crib	loft	O	NOTES	

crib	eave	ridge	L	N	T	F	S	R	C	crib	loft	O	NOTES
5-4	3-0	5-0	6	V	Ax	M	HV	M	F	St	H	—	
5-7	7-0	11-6	5	H	Ax	Q	H	M	P	E	E	—	Continuous primary. Barn fire damaged.
6-0	5-4	5-0	7	V	Ax	—	V	M	F	E	E	—	Moved and rebuilt.
6-0	7-5	12-6	5	H	Ax	Q	H	M	G	E	E	SW	Continuous primary.
5-8	11-0	3-7	8	S	Ax	P	HV	M	F	Ho	H	SW	
8-0	5-0	7-0	6	V	Ax	P	H	M	F	St	T	—	
6-0	5-8	10-6	5	H	Saw	Q	H	M	F	HT	HT	E	Hewn truss members.
6-1	6-6	10-6	5	H	Saw	Q	H	W	F	E	E	—	Hewn purlins. Adjacent house 1856.
6-8	5-10	10-6	6	H	Ax	Q	H	M	F	A	T	SW	
5-0	6-6	12-6	4	H	Ax	Q	H	M	P	E	E	SE	Cribs divided into 2 pens. RH 1841–1923.
6-0	5-8	10-6	7	Sa	Ax	P	V	M	F	St	T	SW	Cribs open to center space.
5-8	7-3	11-3	8	V	Ax	Q	V	M	F	C	H	SE	Cribs open to center. 1938. AS 1896–1974.
6-0	6-9	10-6	7	V	Ax	P	H	M	P	Ho	H	S	
6-5	7-0	14-6	8	V	Ax	Q	H	M	F	Ho	St	SW	N. crib gone. Much of S. crib cut away.
6-4	7-7	10-6	5	H	Ax	Q	H	M	F	St	HT	SW	Cribs open to center. Cribs of unequal size.
5-4	5-10	7-6	5	V	Ax	P	H	M	P	St	St	N	
5-8	5-6	7-6	7	H	Ax	P	V	M	P	E	E	N	Cribs largely cut away.
7-6	6-0	8-0	5	H	Ax	P	V	M	F	St	T	NW	Extensions 2-8. Second floor gone.
—	—	—	—	V	Ax	—	V	M	F	St	H	NW	

Table B1 continued
Information Tables for All Barns Surveyed

Half-Double-Cantilever Barns

	BUILDER	LOCATION	OVERALL DIMENSIONS		WIDTH		DEPTH				CANTILEVER Primary	
			width	depth	cant	crib	space	cant	crib	#	end	mid
BL5	Sam H. Ogle	Law's Chapel Rd.	37-5	26-2	6-0	10-7	10-3	5-10	14-6	7	6x11	6x11
BL45	Caleb Glover	Glover Rd.	52-1	33-3	9-2	12-4	18-3	8-6	16-3	5	7x6	7x6
BL71	Minice	Old Cade's Cove Rd.	34-6	27-8	6-0	10-0	8-5	7-10	12-0	7	8x9	8x10
BL80	Thos. H. McNeilly	Butterfly Loop Rd.	42-9	28-10	6-3	12-3	12-0	8-2	12-6	7	7x8	8x11
BL85	John Riddle	Old Walland Highway	48-0	31-10	8-0	12-0	16-0	8-11	14-0	10	9x7	9x12
BL87	McDonald	U.S. 129	44-0	36-0	8-0	12-0	12-0	9-0	18-0	10	7x6	7x12
BL89	John A. Downey	Lee Shirley Rd.	37-8	30-0	8-0	9-10	10-0	8-0	14-0	5	5x5	5x8
BL91	M.Vanburen Best	Porter Rd.	—	33-8	—	11-8	9-9	8-0	17-8	9	13x9	13x7
BL93	Woodard	Indian Warpath Rd.	49-11	15-11	10-0	13-10	12-3	8-0	15-1	9	10x8	10x8
JO1	———	U.S. 421	58-1	35-7	10-0	15-10	16-3	9-11	15-9	4	12x10	12x10
SE13	Giles Emert	Richardson Cove Rd.	41-10	29-10	8-0	12-0	9-10	8-10	12-2	10	9x10	9x10
SE28	D. C. Robertson	Roberts Schoolhouse	42-2	30-0	8-0	12-0	10-2	9-0	12-0	5	8x6	8x10
SE40	Delozier	U.S. 411	40-3	32-0	10-1	10-0	10-2	10-0	12-0	10	7x9	7x11
SE49	———	Old Sevierville Pike	62-10	34-11	8-10	12-0	20-0	16-11	18-0	8	11x16	11x16
SE71	Trotter (?)	Ogle Drive	41-9	31-11	8-0	11-9	10-2	9-1	13-9	10	5x6	6x10
SE165	William Roberts	Roberts Schoolhouse	42-0	26-6	6-8	12-8	10-0	7-4	11-12	8	7x7	7x7
SE166	Stephen Roberts	Roberts Schoolhouse	39-10	33-0	8-2	10-10	10-0	9-0	15-0	10	9x8	9x14
SE171	Fox	Walden Creek Rd.	44-4	31-10	7-8	11-5	13-10	7-10	16-2	8	7x8	8x9

| DIMENSIONS | | BARN HEIGHTS | | | | | | | | | | USE | | | |
Secondary end	mid	crib	eave	ridge	L	N	T	F	S	R	C	crib	loft	O	Notes
6x7	9x7	5-7	6-6	8-6	5	H	Ax	Q	V	M	F	E	E	NE	SHO 1865-1949.
7x9	7x15	6-5	—	—	5	S	Ax	Q	V	M	P	A	H	E	Pre-1900. CG 1872-1953.
7x9	7x10	5-0	6-6	11-6	5	S	Ax	Q	V	M	F	H	H	NW	Center-opening cribs. c. 1910.
8x8	8x10	6-4	7-2	9-0	5	H	Ax	Q	M	M	F	C	H	—	c. 1910. THMc 1879-1965.
7x6	6x10	6-6	5-6	9-6	5	V	Ax	Q	V	M	G	C	H	—	Extension 2-9. JR 1843-1914.
7x6	7x12	6-6	6-6	6-6	5	H	Ax	P	V	M	P	C	E	S	West crib has 3 pens. Pre-1900.
7x7	7x10	7-2	8-0	12-6	5	H	Ax	Q	H	M	P	E	E	S	Continuous sill. JAD 1844-n.d.
14x9	14x9	6-0	5-9	11-6	5	V	Ax	Q	H	M	P	C	HT	E	Cont. primary. MVB 1837-99.
8x7	9x15	6-9	6-8	13-6	6	V	Ax	Q	H	M	F	St	T	SW	Continuous primary. c. 1880.
11x11	10x12	6-6	6-4	13-0	6	V	Ax	—	V	M	G	St	H	N	2 story cribs.
7x7	7x10	6-4	7-0	12-6	5	H	Ax	Q	H	M	G	Ho	H	SE	GE 1857-1938.
10x8	9x11	7-0	7-6	12-6	6	H	Ax	P	H	M	G	St	T	NE	DCR 1840-n.d.
6x9	7x12	5-8	5-5	7-6	5	H	Ax	P	H	M	F	St	H	SE	Built c. 1890 from log house.
9x11	15x12	8-6	5-7	—	5	H	Ax	P	V	M	F	St	HT	S	N. and w. cantilevers only.
6x6	7x10	6-10	7-1	13-4	6	H	Ax	Q	H	M	G	St	E	N	1870s.
7x7	7x13	6-6	6-0	10-0	5	H	Ax	Q	H	M	P	St	E	N	WR 1854-1923.
8x9	9x11	7-6	6-10	12-6	5	H	Ax	Q	H	M	F	H	E	SE	SR 1855-1907.
9x8	9x8	5-7	7-6	7-6	6	Sa	Ax	Q	H	M	P	C	H	NW	1870s.

Table B1 continued
Information Tables for All Barns Surveyed

Four-Crib Double-Cantilever Barns

	BUILDER	LOCATION	OVERALL DIMENSIONS		WIDTH		DEPTH				CANTILEVER Primary	
			width	depth	cant	crib	space	cant	crib	#	end	mid
BL9	J. H. McCampbell	Old Cades Cove Rd.	56-3	37-11	8-0	14-0	12-2	1-3	12-0	12	11x11	9x10
BL33	Jacob Peters	Miser Station Rd.	61-0	33-5	9-8	15-10	10-0	2-2	11-7	16	7x7	8x8
BL99	Langston Clark	Six Mile Rd.	56-0	33-2	1-10	20-0	12-4	0-10	11-0	10	9x9	—
SE33	———	U.S. 441	40-4	34-0	4-2	12-0	12-0	2-0	10-0	8	8x8	8x10
SE38	———	U.S. 411	39-8	35-9	1-6	12-4	12-0	—	12-2	4	—	10x9
SE96	———	McCleary Rd.	36-11	35-6	1-10	12-0	10-1	2-0	9-9	8	9x7	9x7
SE127	Elijah Henderson	Walnut Grove Rd.	52-0	41-0	1-0	18-0	14-0	1-0	13-0	—	8x12	8x13
SE177	———	New Era Rd.	40-4	31-9	1-11	12-0	12-6	1-0	9-9	8	5x11	6x10

DIMENSIONS Secondary		BARN HEIGHTS										USE			
end	mid	crib	eave	ridge	L	N	T	F	S	R	C	crib	loft	O	Notes
10x8	10x8	8-0	5-6	16-8	6	S	Ax	P	H	M	G	H	H	NE	2nd aisle 11-0. JHMc 1856-1938.
7x8	7x11	6-8	6-0	6-0	7	V	Ax	Q	V	M	P	H	T	W	2nd aisle 5-10. JP 1832-1901.
9x8	—	6-8	6-5	6-5	6	S	Ax	Q	H	M	F	C	H	N	2nd aisle 9-6. c. 1890.
9x8	13x9	6-4	6x3	15-0	6	H	Ax	Q	H	M	F	St	T	NE	2nd aisle 10-0.
—	9x9	6-7	7-6	15-0	6	H	Ax	Q	H	M	P	St	HT	S	2nd aisle 11-7. Extensions 2-0.
9x8	9x8	4-6	8-2	12-0	5	H	Ax	P	V	M	G	C	T	E	2nd aisle 11-9.
8x8	8x7	5-10	6-4	13-8	4	H	Ax	Q	V	M	F	C	H	E	2nd aisle 13-0. EH 1820-98.
6x10	8x11	5-6	6-6	14-6	5	H	Ax	Q	H	M	P	St	T	SE	2nd aisle 10-0.

Table B1 continued
Information Tables for All Barns Surveyed

Unclassifiable Barns

	BUILDER	LOCATION	OVERALL DIMENSIONS		WIDTH		DEPTH				CANTILEVER Primary	
			width	depth	cant	crib	space	cant	crib	#	end	mid
BL39	Mary Cochran	Binfield Rd.	35-2	—	8-5	12-0	—	8-2	19-10	10	7x5	7x10
BL53	Ira W. Peery	Bethlehem Rd.	54-1	39-7	—	—	—	—	—	10	9x9	9x14
SE1	———	U.S. 441 at Seymour	31-9	18-10	13-9	9-4	8-10	10-0	—	5	—	—
SE57	Jacob H. Seaton	Newcomb Hollow Rd.	30-0	18-0	9-0	12-0	—	—	18-0	5	—	—
SE140	———	Maple Branch Rd.	23-8	18-5	7-8	16-0	—	8-6	9-11	7	8x6	8x6

DIMENSIONS Secondary		BARN HEIGHTS										USE			
end	mid	crib	eave	ridge	L	N	T	F	S	R	C	crib	loft	O	Notes
6x6	7x11	5-8	7-6	12-0	5	H	Ax	Q	H	M	F	St	T	NE	c. 1885. Three crib barn.
11x8	11x8	8-2	7-2	15-0	6	H	Ax	Q	H	M	F	H	HT	W	Three crib barn. IWP 1884-1978.
6x5	6x5	5-6	10-0	5-0	6	S	Ax	P	V	M	P	E	E	SW	
10x7	7x11	6-6	6-3	8-0	6	H	Ax	P	H	M	P	E	E	SE	Single crib single cantilever.
6x5	6x5	3-10	7-2	6-0	5	H	Ax	—	V	M	P	St	E	NW	1 crib. NW and NE cantilevers.

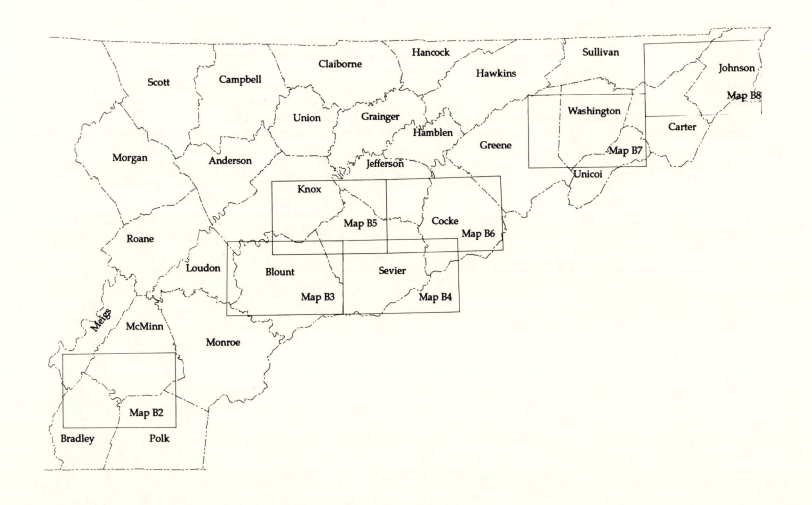

Map B1. *State of Tennessee, keyed to maps B2 through B8.*

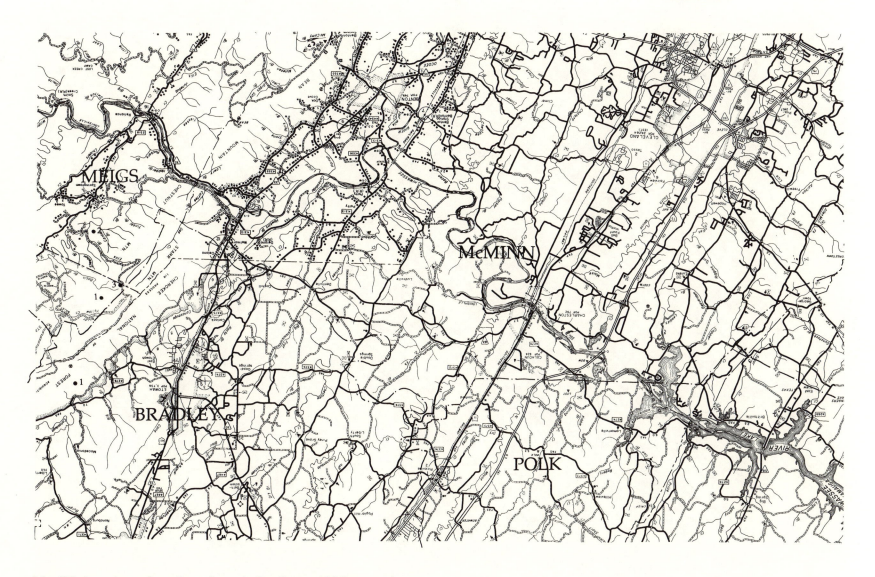

Map B2. *Locations of cantilever barns in Bradley and Meigs counties.*

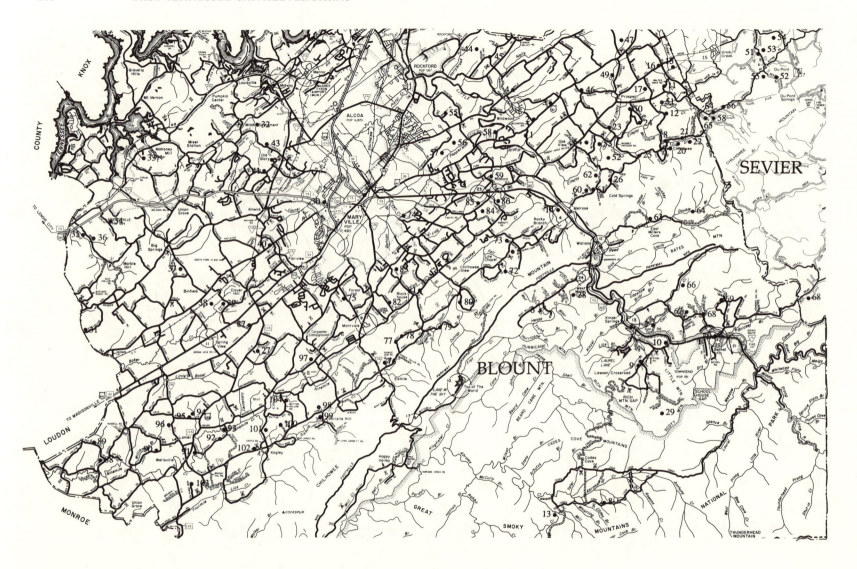

Map B3. *Locations of cantilever barns in south Blount and west Sevier counties.*

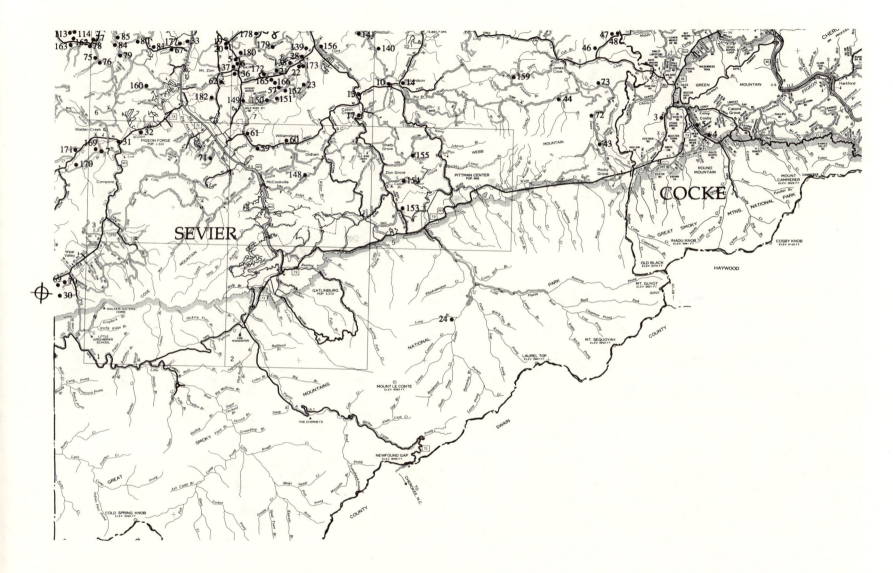

Map B4. *Locations of cantilever barns in south Sevier and south Cocke counties.*

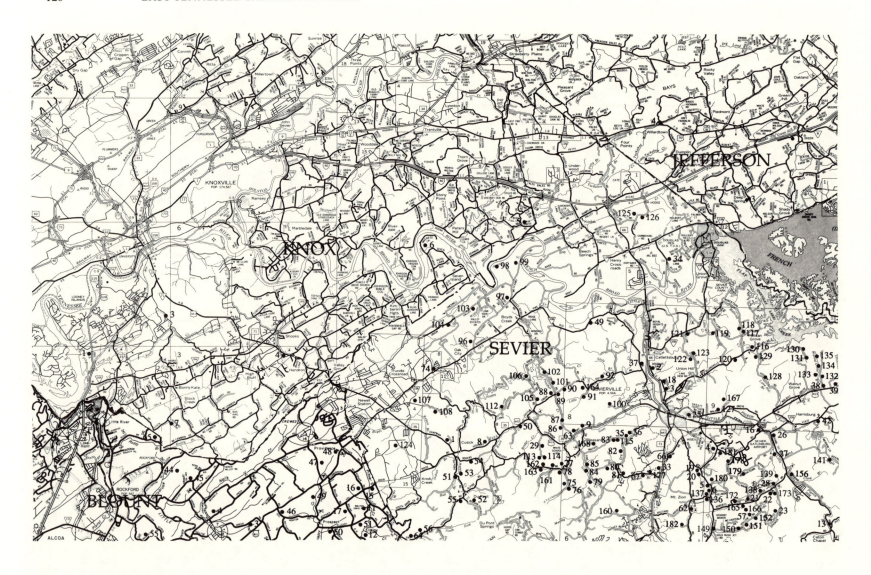

Map B5. *Locations of cantilever barns in south Knox, north Blount, and north Sevier counties.*

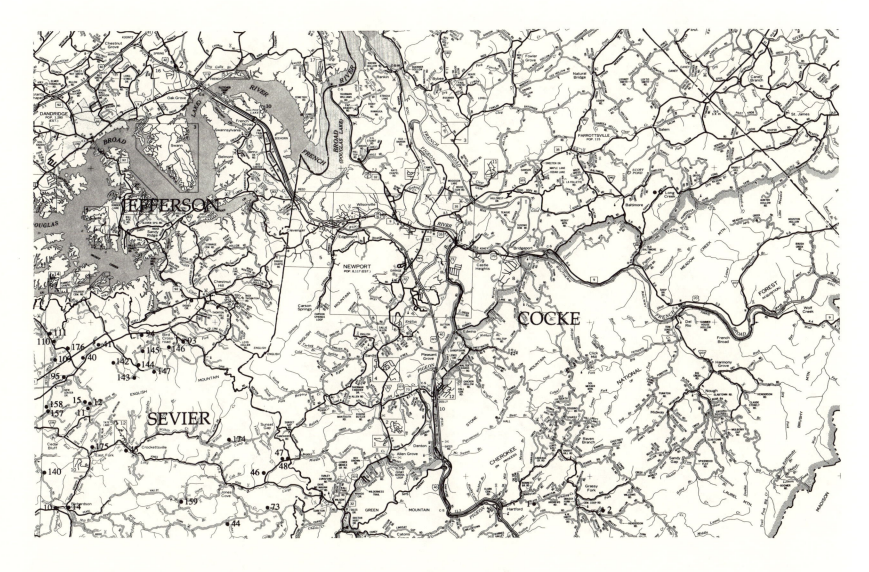

Map B6. *Locations of cantilever barns in north Sevier, north Cocke, and Jefferson counties.*

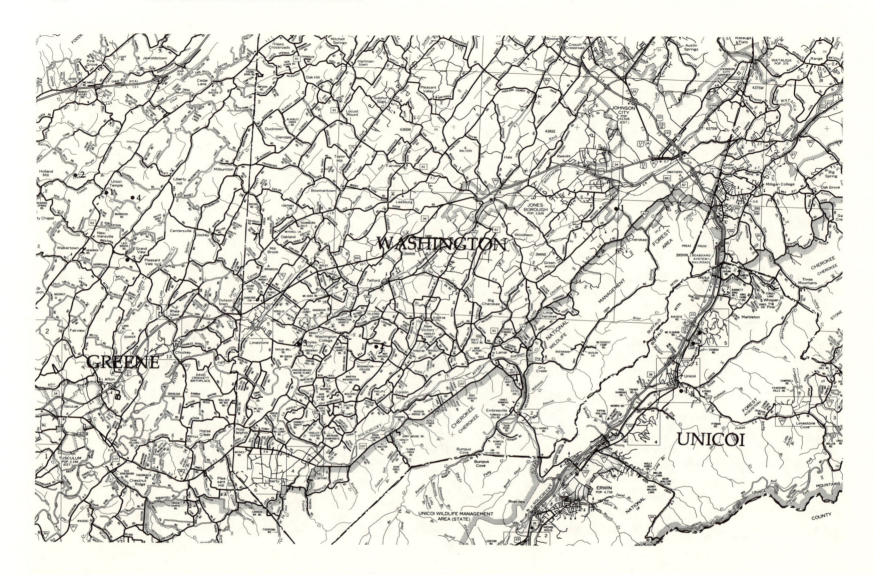

Map B7. *Locations of cantilever barns in Greene, Washington, and Unicoi counties.*

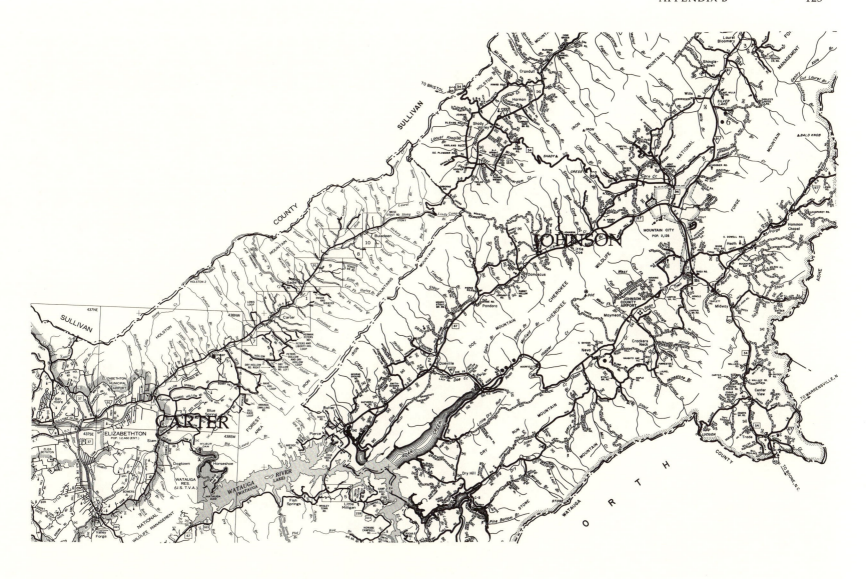

Map B8. *Locations of cantilever barns in Johnson and Carter counties.*

NOTES

Chapter 1
Cantilever Barns Described

1. Blount County is named for William Blount, an early Tennessee land speculator and the first Governor of the United States Territory South of the River Ohio. His name is pronounced *Blunt*. Sevier County is named for John Sevier, a celebrated British and Indian fighter who served as governor of both the State of Franklin and the State of Tennessee. His name is pronounced *Severe*.

2. Several examples are located in Botetourt County, and one has been moved to the Frontier Culture Museum of Virginia in Staunton. See also Peter Letchner, "The Breaks, Virginia," *Pioneer America* 4, no. 2 (July 1972): 1-7.

3. Charles Martin at Alice Lloyd College in Pippa Passes, Kentucky, shared information and photographs of the Caudill barn in Knott County, Kentucky. The presence of a few cantilever barns in the southeastern corner of the state is mentioned but not illustrated in William Lynwood Montell and Michael Lynn Morse, *Kentucky Folk Architecture* (Lexington: Univ. Press of Kentucky, 1976), 65-66.

4. John Burrison discovered a single crib barn with cantilevered overhangs on three sides in Dawson County on Route 9E near the Lumpkin campground.

5. John Morgan and Ashby Lynch, Jr., "The Log Barns of Blount County, Tennessee," *Tennessee Anthropologist* 9, no. 2 (1984): 90.

6. Although our fieldwork concentrated on cantilever barns, we did come across widely scattered examples of log barns without cantilevers. In Sevier County, we noted fewer than six two- or four-crib log barns without cantilevers. Log cabins and particularly log smokehouses are more common.

7. This distribution of corner timbering types concurs with that observed by Fred B. Kniffen and Henry Glassie, "Building in Wood in the Eastern United States: A Time-Place Perspective," *Geographical Review* 56 (1966): 59. Our field evidence, like that of John Morgan on log houses in Blount, Grainger, and Morgan counties, does not support theories advanced by Jordan and Roberts regarding relationships between corner notching and the species of wood employed. See John T. Morgan, *The Log House in East Tennessee*, (Knoxville: Univ. of Tennessee Press, 1990), 34-42.

8. Joseph Buckner Killebrew, *Tennessee: Its Agricultural Resources and Mineral Wealth*, (Nashville: Tavel, Eastman & Howell, 1874), 51. By the 1880s, steam-powered threshing machines came into use in the region, which would have made it possible to cease hand-threshing, so for over a century a threshing floor would have been unnecessary. In *On Horseback: A Tour in Virginia, North Carolina, and Tennessee*, (Boston: Houghton Mifflin, 1888), Charles Dudley Warner records an encounter with such a machine in Greene County, and the authors heard anecdotal accounts of late nineteenth-century threshing machines from several barn owners.

9. Glassie, "The Old Barns of Appalachia," 29.

10. Dendrochronology involves matching patterns of growth rings from woods which can then be compared against samples from trees of the same species having known dates. Most of the timber in cantilever barns came

from pine trees approximately fifty years old, and it cannot be compared to hardwood specimens, even those taken from trees in the general region. Pine trees lack longevity, and we have been unable to locate specimen samples suitable for comparative analysis. We are indebted to Dr. Frank Woods of the Department of Forestry, Fisheries, and Wildlife at the University of Tennessee, Knoxville, and Betsy Groton of the Forestry Division, Tennessee Valley Authority, for their assistance with dendrochronology.

11. A historical marker placed on the Jesse Atchley Barn (Sevier 151) indicates that it was built around 1800; the barn was relocated to its present location and inexpertly restored in the 1970s, and it is unclear whether the 1800 date is accurate.

12. Milton B. Newton, Jr., and Linda Pulliam-DiNapoli, "Log Houses as Public Occasions: A Historic Theory," *Annals of the Association of American Geographers* 67 (1977): 360-83.

13. Pages of *The American Agriculturist* include barn plans from time to time. A book, *Barn Plans and Out-Buildings* by Byron D. Halsted (New York: Orange Judd, 1881), gathers many of these together in one volume.

14. John C. Campbell, *The Southern Highlander and His Homeland* (New York: Russell Sage Foundation, 1921; rpt. Lexington: Univ. Press of Kentucky, 1969), 143. This book also has a photograph of a "working" in progress.

Chapter 2
Origins of the Cantilever Barn

1. See Robert J. Naismith, *Buildings of the Scottish Countryside* (London: Victor Gollancz Ltd., 1985) and John B. Rehder, "The Scotch-Irish and English in Appalachia," in *To Build in a New Land*, ed. Allen G. Noble (Baltimore: Johns Hopkins Univ. Press, 1992), 95-118.

2. Splendid illustrations of European log building details may be seen in Hermann Phleps, *The Craft of Log Building: A Handbook of Craftsmanship in Wood*, trans. Roger MacGregor, (1942; rpt. Ottawa: Lee Valley Tools, Ltd., 1982). In comparison with Phleps's examples, American log buildings are very simply constructed.

3. Sergei Ivanovich Rudenko, *Frozen Tombs of Siberia: The Pazyryk Burials of Iron Age Horsemen*, trans. M. W. Thompson (Berkeley: Univ. of California Press, 1970).

4. Terry G. Jordan, *American Log Buildings: An Old World Heritage,* (Chapel Hill: Univ. of North Carolina Press, 1985), 10, 146. C. A. Weslager first proposed the Finno-Swedish origins for log building in *The Log Cabin in America: From Pioneer Days to the Present* (New Bruswick, N.J.: Rutgers Univ. Press, 1969). More recently, the case for Finno-Swedish transmittal to the New World has been extensively documented in Terry G. Jordan and Matti Kaups, *The American Backwoods Frontier: An Ethnic and Ecological Interpretation* (Baltimore: Johns Hopkins Univ. Press, 1989). Not all scholars support this conclusion, however. Bernard L. Herman, for example, in *Architecture and Rural Life in Central Delaware, 1700-1900* (Knoxville: Univ. of Tennessee Press, 1987), proposes that American log construction should be seen as a development from pan-European log building traditions.

5. Jordan, *American Log Buildings,* 58. See also Fred Kniffen, "On Corner Timbering," *Pioneer America* 1, no. 1 (Jan. 1969): 1-8.

6. From studies of Indiana log cabins, Warren Roberts has illustrated the wide variety of tools available in the American back country. These were more plentiful than might at first be imagined and were used by their owners in Indiana in return for produce or as a means of barter. See Warren Roberts, "The Tools Used in Building Log Houses in Indiana," *Pioneer America* 9, no. 1 (1977): 32-61.

7. Henry Glassie, "The Appalachian Log Cabin," *Mountain Life and Work* 39, no. 4 (Winter 1963): 5-14.

8. Jordan, *American Log Buildings,* 14-15. No evidence has been found to indicate that the cantilever barns were the work of the "semiprofessional, often itinerant carpenters" mentioned by Jordan.

9. Joseph Buckner Killebrew, *Introduction to the Resources of Tennessee* (Nashville: Tavel, Eastman & Howell, 1874), 438.

10. A photograph of a log house under construction was included in Campbell, *The Southern Highlander,* and a complete demonstration of log building was provided by two elderly men from the Georgia mountains for Eliot Wiggenton, ed., *The Foxfire Book* (New York: Doubleday, 1972), 53-107.

11. Morgan, *The Log House in East Tennessee,* 98-107.

12. Max Geschwend, *Schweizer Bauernhäuser: Material, Konstruktion und*

Einteilung (Bern: Verlag Paul Haupt, 1983) illustrates Swiss examples and details of the log techniques used there. Examples of these buildings are preserved at the Ballenberg open-air museum near Brienz.

13. Phleps, *The Craft of Log Building*, 306-22.

14. Robert F. Ensminger, "A Search for the Origin of the Pennsylvania Barn," *Pennsylvania Folklife* 30, no. 2 (Winter 1980-81): 66-67.

15. Ensminger, 58. Well-preserved examples may be seen in the Rheinland open-air museum at Kommern.

16. Terry G. Jordan, "Alpine, Alemannic, and American Log Architecture," *Annals of the Association of American Geographers*, 70, no. 2 (1980): 168. See also Terry G. Jordan, "Some Neglected Swiss Literature on the Forebay Bank Barn," *Pennsylvania Folklife* 37, no. 2 (1987-88): 75-80.

17. Henry Glassie, *Patterns in the Material Folk Culture of the Eastern United States* (Philadelphia: Univ. of Pennsylvania Press, 1968), 55.

18. Allen G. Noble, *Wood, Brick, and Stone: The North American Settlement Landscape. Volume 2: Barns and Farm Structures* (Amherst: Univ. of Massachusetts Press, 1984), 2.

19. Fred B. Kniffen, "Folk Housing: Key to Diffusion," *Annals of the Association of American Geographers* 55, no. 4 (1965): 549-77.

20. Henry Glassie, "The Pennsylvania Barn in the South," 18-19.

21. Walter M. Kollmorgen, "Observations on Cultural Islands in Terms of Tennessee Agriculture," *East Tennessee Historical Society's Publications* 16 (1944): 68. Of these Pennsylvania Germans, the author notes that "somehow they differed from many other frontiersmen as long as they maintained community integrity. By and large they were small farmers and stuck to diversified farming."

22. We are indebted to Gary Hamilton, a photographer with the Knoxville *Journal,* for providing information and photographs of the barn.

23. Information on the Knox County bank barns comes from the county historical inventory conducted in 1984-85. The bank barns are inventory numbers 3788B, 6947, 7358, and 7379. Copies of the field reports are on microfilm in the East Tennessee Historical Center, Knoxville.

24. See Alexander Opolovnikov and Yelena Opolovnikova, *The Wooden Architecture of Russia: Houses, Fortifications, Churches* (New York: Harry N. Abrams, 1989), 83-139.

25. Raymond Evans, "Fort Marr Blockhouse, Last Evidence of America's First Concentration Camp," *Journal of Cherokee Studies* 2 (Spring 1977): 256-63.

26. Nancy L. O'Neil, "The Swaggerty Blockhouse," *Smoky Mountain Historical Society Newsletter* 11, no. 1 (Spring 1985): 4-7. According to family records, it was built by James Swaggerty, who was then only fifteen years old. More probably the blockhouse was the work of James's father, Frederick, or uncle Abraham. A veteran of the Revolutionary War, Abraham had moved to Tennessee from Pennsylvania, receiving on the last day of 1786 a land grant, which was registered in October of the following year. Church records report troubles with Indians during the period from 1787 to 1789, necessitating the construction of a fort. Adequate provision against attack was afforded by its sturdy construction, overhanging upper floor, and location over a flowing spring. The Swaggerty blockhouse is listed on the National Register of Historic Places.

27. Eugen Adolf Hermann Petersen and Felix von Luschan, *Reisen im Südwestlichen Kleinasien* (Vienna, 1899), 147-48. The purpose of the Anatolian log house is not given in the text; one suspects from its construction and placement on an earthen mound that the building was intended as a lookout. Roof planks are supported on tapered cantilevers overhanging all four sides of the cubical house and supporting a sod roof weighted down with large stones.

28. Jordan, *American Log Buildings*, 75.

Chapter 3
The Builders

1. François André Michaux, *Travels to the Westward of the Allegany Mountains in the States of Ohio, Kentucky and Tennessea* (London: B. Crosby, 1805), 224.

2. See Stanley J. Folmsbee, Robert E. Corlew, and Enoch L. Mitchell, *Tennessee: A Short History* (Knoxville: Univ. of Tennessee Press, 1969), 48-90.

3. Campbell, *The Southern Highlander*, provides an excellent description of settlement patterns in the Appalachians. For an individual family history of the move into Appalachia, see John Egerton, *Generations: An American Family* (New York: Simon and Schuster, 1986). David H. Fischer's study, *Albion's Seed: Four British Folkways in America* (New York: Oxford Univ.

Press, 1989), places the Scots-Irish in the cultural perspective of overall British immigration.

4. Killebrew, *Tennessee: Its Agricultural Resources,* 38-39.

5. Microfilm copies of the original census rolls for Blount and Sevier counties were consulted for all census information. In footnotes, these will be referenced by census type, year, and other identifying information. Pagination of the original rolls is often erratic. Sevier County cemeteries were extensively inventoried by the Smoky Mountain Historical Society and published as *In the Shadow of the Smokies: Sevier County Cemeteries* (Sevierville, Tenn.: Smoky Mountain Historical Society, 1984). Blount County inscriptions were taken from Edith B. Little, *Blount County Tennessee Cemetery Records* (Evansville, Ind.: Whipporwill Publications, [1981]).

6. Killebrew, *Introduction to the Resources of Tennessee,* 433.

7. Pryor Duggan's records come from the population censuses of 1860 (District 6, page 325), 1870 (District 6, family 89), and 1880 (District 6, family 5) and the agricultural censuses of 1860 (District 6, page 1, line 35), 1870 (Subdivision 27, page 5, line 6), and 1880 (District 6, page 1, line 5). Wilson Duggan's records come from the population census of 1840 (page 173) and 1850 (page 198 or 797) and the agricultural census of 1860 (District 3, page 13, line 23).

8. "Sevier County Members of Tennessee General Assembly," *Sevier County News-Record,* July 1, 1976.

9. John Andes's records come from the population censuses of 1830 (page 92), 1840 (page 166), 1850 (family 379), 1860 (District 6, page 325), 1870 (District 6, family 81), and 1880 (District 6, family 183) and the agricultural censuses of 1860 (District 6, page 1, line 39), 1870 (Subdivision 27, page 5, line 1) and 1880 (District 6 Walden Creek, page 14, line 9). His name in the 1850 population census is given as *Anders,* and in 1870 he is listed under *Andrews.*

10. Data on Riley H. Andes come from the population censuses of 1870 (District 7, family 118) and 1880 (District 5, family 4) and the agricultural censuses of 1870 (District 7, page 9, line 27) and 1880 (District 5, page 1, line 3). Data on John W. Andes come from the population censuses of 1870 (District 7, family 119) and 1880 (District 5, family 1) and the agricultural censuses of 1870 (District 7, page 9, line 28) and 1880 (District 5, page 1, line 1).

11. "Sevier County Members of Tennessee General Assembly."

12. The authors are indebted to Beulah Linn's research for "A History of the Bank of Sevierville," (1988), the source of personal information about Will Catlett.

13. William Catlett's records come from the population censuses of 1850 (page 503), 1860 (District 5, family 607), 1870 (District 5, family 7), and 1880 (District 5, City of Sevierville, family 224) and the agricultural censuses of 1850 (Subdivision 12, page 569, line 9), 1860 (District 5, page 33, line 23), 1870 (District 5, page 1, line 3), and 1880 (District 5, page 6, line 8).

14. Abraham McMahan's records come from the population census of 1870 (District 5, family 83) and the agricultural censuses of 1850 (Subdivision 12, page 577, line 9) and 1860 (District 5, page 35, line 18). George W. McMahan is listed in the 1880 population census (District 5, family 6) and 1880 agricultural census (District 5, page 1, line 5).

15. Through his wife, McMahan had connections to the barn builders of Middle Creek considered in chapter 5. Marusia was a daughter of George M. and Sarah Yett Henderson, and her brother Robert Henderson built Barn 138. See figure 50.

16. Richard Reagan's records come from the population censuses of 1830 (page 90), 1840 (page 185), 1850 (page 843 or 891), 1860 (District 10, family 73), and 1880 (District 10, family 213), and the agricultural censuses of 1850 (Subdivision 12, page 585, line 21), 1860 (District 10, page 9, line 18), 1870 (Subdivision 27, page 17, line 40), and 1880 (District 10, page 13, line 9).

17. The local community eventually became known as Reagan Town because of the numerous Reagan descendants. See Veta King, "Dupont's History Started at Line Where Dew Won't Form," Knoxville *News-Sentinel South,* Mar. 20, 1991.

18. Present owners did not know the first names of the Sharps who are credited with building the barns. Beulah Linn has informed us that Willis B. Sharp married Mary Zollinger, the daughter of Alexander Zollinger, a Revolutionary War veteran from Virginia. "Willis left Mary when the children were young and they lived with her father until he died in the 1850s—ca. 1852" (Beulah Linn, letter to the authors, Mar. 25, 1991). It seems likely that the Sharp children built the barns when they were grown, but we cannot ascertain this.

19. Matthew Tarwater's records come from the population censuses of 1870 (District 10, family 157), 1880 (District 10, family 42), and 1900 (District 10, family 291) and the agricultural censuses of 1860 (District 10, page 1, line

1), 1870 (Subdivision 27, page 17, line 2), and 1880 (District 10, page 3, and line 9). James R. Tarwater's records come from the 1880 population census (District 10, family 35) and the 1880 agricultural census (District 10, page 3, line 5). William D. Tarwater's records come from the population censuses of 1880 (District 5, family 43) and 1900 (District 10, family 289) and the 1880 agricultural census (District 5, page 4, line 2). Additional information about family names and dates comes from Pollyanna Creekmore, "Matthew Tarwater Family Bible," *Smoky Mountain Historical Society Newsletter* 14, no. 4 (Winter 1986): 114-15.

20. Carroll Van West, *Tennessee Agriculture: A Century Farm Perspective* (Nashville: Tennessee Dept. of Agriculture, 1986), 105.

21. Present barn owners seldom knew specific dates of their barns. If they supplied a date at all, the barns was generally said to be 100 years old, and we suspect that the same answer might well have been given in 1975 or in 1995. No one reported owning a 90- or 110-year-old barn.

22. Nora Harbin DeArmond, "Bits and Pieces of History about Smoky Mountain Foothills People: Ingle Families," *Smoky Mountain Historical Society Newsletter* 16, no. 2 (Winter 1990): 98-100.

23. James Bohanon's records come from the population censuses of 1870 (District 13, family 3) and 1880 (District 13, family 115) and the agricultural censuses of 1870 (page 31, line 39) and 1880 (District 13, page 22, line 10). His grave marker in the Williamsburg Cemetery is the source of his wife's first name. Beulah Linn has supplied additional personal information about the Bohanons.

24. Interestingly enough, the family became patrons of education: James and Clarinda donated land across the road from their barn for the construction of the Williamsburg School (built 1896). The building still stands, although used now as a Pentecostal church. The Bohanons also gave land for the Williamsburg Cemetery, where both are buried. (Beulah Linn, letter to the authors, Mar. 25, 1991.) Mrs. Linn's mother knew Bohanon ("Uncle Jim") as a highly respected man in the community.

25. Whitehead family records come from the population censuses of 1870 (District 3, family 172), 1880 (District 3, family 40), and 1900 (District 3, families 37 and 38) and the agricultural census of 1880 (District 3, page 4, line 4). His father appears in the 1850 agricultural census (Subdivision 12, page 551, line 35). Information on other nineteenth-century Whiteheads in Sevier and Knox counties is presented in Albert W. Dockter, Jr., "Zwinglites in Sevier and Knox Counties?????," *Smoky Mountain Historical Society Newsletter* 15, no. 1 (Spring 1989): 10-13. How these Whiteheads connect to the barn builder is unclear, as the names and dates do not coincide.

26. It is not uncommon in the census records for Sevier County to note eleven-year-olds who, although attending school, are still illiterate. Common schooling was available only to those who could afford tuition, and even so the school year lasted only four to five months a year. A single teacher managed each school, teaching pupils of all levels in one room. The Social Survey of Blount County undertaken in 1930 by students at Maryville College observed the decline there in one-teacher schools, from fifty-six in 1919-20 to only thirty-one in 1928. More telling is the observation that in 1919 there were only three college graduates teaching elementary school and none in the high schools; by 1928, the number schoolteachers who were college graduates had risen to fifty-six. Social statistics from the 1860 census indicate that Sevier County then had forty-two common schools, but a long-time supporter of education, Daniel G. Emert, observed in 1900 that "half a century ago there were few people under fifteen years of age who could read or write. Now there are few who cannot read or write." See Beulah Duggan Linn, "50 Year History of Educational Growth," *Sevier County News-Record,* Jan. 15, 1976.

27. It is possible that John C. and not John W. built the Whitehead barn. The adjacent house is dated 1895.

Chapter 4
Regional Factors

1. Earl C. Case, *The Valley of East Tennessee: The Adjustment of Industry to Natural Environment* (Nashville: State Geological Survey, 1925), 7.

2. Of the forty-three types of termites in the United States, only two subterranean types, the *Reticulitermes flavipes* and *Reticulitermes virginicus,* inhabit the humid, damp, and warm conditions of the American Southeast. Termites in northern portions tend to live in buildings, not in the ground. Southern termites have thin skins, even for termites, and in less-humid conditions, they lose body water and die. Even though they invade wood by constructing shelter tubes, usually of mud, "periodically they must

return to the moist galleries in the soil to replenish the water lost from their bodies in the relatively dry air of their workings above ground." Harry B. Moore, *Wood-Inhabiting Insects in Houses: Their Identification, Biology, Prevention and Control* (Washington: U.S. Dept. of Agriculture, 1979), 14-21.

3. Joseph Buckner Killebrew, *Tobacco: Its Culture in Tennessee,* (Nashville: Tavel, Eastman & Howell, 1876), 48-9.

4. Donald W. Buckwalter, "Effects of Early Nineteenth Century Transportation Disadvantage on the Agriculture of East Tennessee," *Southeastern Geographer* 27 (1987): 18-37.

5. Killebrew, *Introduction to the Resources of Tennessee,* 356.

6. For two examples of these accounts, see Charles D. Warner, *On Horseback: A Tour in Virginia, North Carolina, and Tennessee* (Boston: Houghton Mifflin, 1888) and David Hunter Strother, *The Old South Illustrated, by Ponte Crayon,* ed. Cecil D. Eby, Jr., (Chapel Hill: Univ. of North Carolina Press, 1959).

7. Margaret Ann Roth, "The End of Isolation: Transportation and Communication," in *The Gentle Winds of Change: A History of Sevier County 1900-1930* (Maryville, Tenn.: Smoky Mountain Historical Society, 1986), 54-56.

8. For an example of this negative characterization, see "Corn Planting," *The American Agriculturist* 32 (Apr. 1873): 139-40.

Chapter 5
The Barns of Middle Creek

1. Detailed agricultural census data is available for the years 1850 through 1880. For these years, farm data are given by the name of each farmer. The 1890 census was destroyed by fire; and from 1900 on, the Bureau of the Census has released aggregate data but not maintained the records from individual farms which would have been of such importance in this study and to other inquiring historians.

2. John Arthur Shields, quoted in Beulah Duggan Linn, "Joshua Tipton, Forgotten Revolutionary Soldier," *Sevier County News-Record,* July 22, 1976. The authors are grateful to Mrs. Linn for sharing this information with them.

3. Smoky Mountain Historical Society, *Sevier County Cemeteries,* n.d., 280. The papers of Isaac Trotter preserved in the Special Collections Library at the University of Tennessee include an undated sketch plat for the Middle Creek campground.

4. The civil districts in Sevier County are subdivisions made for the purposes of census taking and local government elections.

5. A modern map with the land grants drawn over present features was made by Elaine Wells, a member of the Smoky Mountain Historical Society. Mrs. Wells's husband was a descendent of Andrew Wells, a Revolutionary War soldier who settled in Sevier County. Beulah Linn shared her manuscript copy of this map with the authors.

6. "Some old timers on Middle Creek remember their parents talking about a big log house within sight of Henry and Horatio Butler cabins [i.e., near the site of Middle Creek Methodist Church]. Even in the childhood of the parents the roof had fallen in." (Beulah Linn, letter to the authors, Mar. 25, 1991.) Could this building have been the remains of Shields's fort?

7. John Trotter's records come from the 1850 census of agriculture (page 567, line 22) and the population censuses of 1830 (page 112), 1840 (page 164), and 1850 (family 474).

8. Horatio Butler, born in Virginia, moved to Sevier County about 1816. His father, Henry Butler, built a log house which still stands across Middle Creek Road from Henry's log house and barn. Information on Horatio Butler comes from population censuses of 1830 (page 100), 1840 (page 164), and 1850 (family 476). In the 1850 agricultural census, the entry is under Mary T. Butler (page 567, line 24). Further information on the Butler family comes from Beulah Linn, "Henry Butler of Sevier County, Tennessee," *Smoky Mountain Historical Society Newsletter* 8 no. 1 (Spring 1982), 2-3. Mrs. Linn speculates further that the house, which has been dated to ca. 1818, might have been a Shields cabin built outside the fort. (letter to the authors, Mar. 4, 1991).

9. Information on W. H. Trotter comes from the agricultural censuses of 1850 (page 569, line 2), 1860 (District 4, page 17, line 27), 1870 (Subdivision 27, page 27, line 6), and 1880 (District 4, page 2, line 8) and the population censuses of 1850 (family 497), 1860 (District 4, family 503), 1870 (District 4, family 13) and 1880 (District 4, family 22).

10. Information on Isaac Trotter comes from the agricultural censuses of 1850 (page 569, line 3), 1860 (District 4, page 27, line 26), and 1870 (Subdivision 27, page 27, line 7) and the population censuses of 1950 (family 496), 1860 (District 4, family 502), 1870 (District 4, family 14) and 1880 (District 4, family 21). His widow, Mary Trotter, is listed in the 1900 population census

(District 4, family 98), and his son, Amos C. F. Trotter, appears as head of household in the 1910 census (District 4, family 218).

11. Information provided by Beulah Linn, letter to the authors, Sept. 1, 1987.

12. Two justices of the peace were elected for four-year terms from each of the county's seventeen civil districts. These justices comprised the county court, which met quarterly. "In his district the justice of the peace was supreme. People considered a good justice to be one who 'got a lot for his district' in terms of roadwork, bridge building and other things and who voted consistently against tax increases. . . . In addition to his legislative and administrative duties, the justice of the peace . . . could perform marriages and serve as a judge for cases involving minor offenses. He could fine or jail a guilty person but could not sentence anyone to the penitentiary." See William Bruce Wheeler, "Government and Politics" in *The Gentle Winds of Change,* 220.

13. Beulah Linn, letter to the authors, Sept. 1, 1987, notes, "The story is that he [W. H. Trotter] and Frederick Emert made a trip to Georgia to select a house plan" for the residence. The hip-roof house has a double-pile plan—two approximately square rooms on either side of a central hall and staircase. It is unlike other houses in the vicinity, but few frame houses of its period survive. The house (but not the barn) is listed on the National Register of Historic Places.

14. When Trotter's estate was probated, one of his most valuable possessions, aside from livestock, was a set of electromagnetic medical instruments. These were purchased at the estate sale by a physician from another section of the county. It is one of the ironies of record keeping that we know the purchaser, article, and amount paid for even the smallest household implements at this sale, yet there are no clear records about the nature, age, or condition of the house and outbuildings of the Trotter farm.

15. Information on Robert Marshall comes from the agricultural censuses of 1860 (page 25, line 29), 1870 (District 4, page 27, line 4), and 1880 (page 4, line 10) and the population censuses of 1860 (District 4, family 505), 1870 (District 4, family 144), and 1880 (District 4, family 52).

16. The log schoolhouse is now gone; the present church building dates from 1901 (information provided by Beulah Linn in a letter to the authors, Sept. 1, 1987). Information on William Phillip Roberts comes from the agricultural censuses of 1850 (Subdivision 12, page 565, line 17), 1860 (District 4, page 27,

line 11), and 1870 (Subdivision 27, page 27, line 24) and the population censuses of 1860 (District 4, family 528) and 1870 (District 4, family 30).

17. This was a health resort "which catered to Knoxvillians who came to 'take the waters' far away from the rapidly-industrializing city" (Aileen Whaley Fowler, "Fun, Friends and Fellowship: Recreation and Amusements," in *The Gentle Winds of Change,* 196-97). Information on James H. Seaton comes from the agricultural censuses of 1870 (Subdivision 27, page 27, line 22) and 1880 (District 4, page 3, line 9) and the population censuses of 1870 (District 4, family 27), 1880 (District 4, family 65), 1900 (District 4, family 215), and 1910 (District 4, family 167).

18. Information on John A. Seaton comes from the agricultural census of 1880 (District 4, page 3, line 4) and the population censuses of 1870 (District 4, family 51), 1880 (District 4, family 30), 1900 (District 4, family 168), and 1910 (District 4, family 207).

19. Information on Robert Henderson comes from the agricultural censuses of 1870 (Subdivision 27, page 29, line 14) and 1880 (District 4, page 10, line 8) and the population censuses of 1870 (District 4, family 77), 1900 (District 4, family 216), and 1910 (District 4, family 86).

20. Cleason Robertson's full name was Dio Cleason Robertson, the name also of his father (1804-1891). He was often known as D. C., and his name is sometimes spelled as Cleson. In the agricultural census data, it is difficult to tell which man, father or son, is being inventoried, as no ages are provided for the farmers listed. The death date of Dio Cleason the son is unknown. A farmer thought to be the son is listed in the 1880 agricultural census as D. Robinson (District 4, page 11, line 1).

21. Information on William M. Roberts comes from the population censuses of 1900 (District 4, family 208) and 1910 (District 4, family 205). In 1900 William was living with his widowed mother, but by 1910, he was married with four children. Information on Stephen H. Roberts comes from the population censuses of 1900 (District 4, family 210) and 1910 (District 4, family 206), when it is his widow, Mary J. Roberts, who is listed.

22. Information on W. W. Webb comes from the agricultural censuses of 1860 (District 4, page 27, line 9), 1870 (Subdivision 27, page 27, line 21), and 1880 (District 4, page 5, line 9) and the population censuses of 1860 (District 4, family 526), 1870 (District 4, family 26), 1880 (District 4, family 62), and 1900 (District 4, family 166).

23. Information on John Ogle comes from the 1880 agricultural census (District 4, page 4, line 9) and the population censuses of 1900 (District 4, family 205) and 1910 (District 4, family 204).

24. Information on Jacob H. Seaton comes from the 1880 agricultural census (District 4, page 6, line 1) and the population census of 1900 (District 4, family 202).

25. Information on Henry Butler (Jr.) comes from the agricultural censuses of 1850 (page 567, line 28), 1860 (District 4, page 25, line 17), 1870 (Subdivision 27, page 27, line 27), and 1880 (District 4, page 6, line 5).

26. If indeed William J. Trotter was a builder of cantilever barns, he was familiar with the type from both sides of his family. His parents were John M. Trotter and Tryphena Flinn. John M. was a son of John Sevier Trotter, himself a son of John Trotter of Sevier 5. Tryphena was the daughter of George Flinn and Barbara Roberts, a sister to George, Aaron, and William Phillip Roberts of Sevier 173. Thus the builders of the two oldest barns in the Middle Creek community are related through William's family connections (information from Beulah Linn, letter to the authors, Mar. 25, 1991).

27. Beulah Linn, letter to the authors, Mar. 25, 1991.

28. Information on William Sutton comes from the population censuses of 1900 (District 4, family 137) and 1910 (District 4, family 232). In 1900, Martha Roberts, Ellen's mother, is listed as head of household, with William, Ellen, and their five children living with her. That same census notes that William is illiterate.

29. Information on Robert M. Rambo provided by Beulah Linn, letter to the authors, Mar. 25, 1991. Part of the Marshall farm now belongs to Dollywood, an amusement park. See also Beverly Nelson Rambo, *The Rambo Family Tree: Descendants of Peter Gunnarson Rambo, 1611-1986* (Decorah, Iowa: Anundsen, 1986), 165 and 292. According to this book, the Peter Rambo of Tennessee was a sixth-generation descendent of the original Peter Rambo who came to New Sweden.

Chapter 6
The Barns as Vernacular Expression

1. See Smoky Mountain Historical Society, *The Gentle Winds of Change.*

2. See Thomas C. Hubka, *Big House, Little House, Back House, Barn: The Connected Farms Buildings of New England* (Hanover, N.H.: Univ. Press of New England, 1984).

3. The late nineteenth- and early twentieth-century circular-plan barns of Iowa have been inventoried by Lowell J. Soike in *Without Right Angles: The Round Barns of Iowa* (Des Moines: Iowa State Office of Historic Preservation, 1982). Similar barns may be found as far west as the state of Washington, where ten round or polygonal barns survive. The T. A. Leonard Barn in Whitman County was measured and drawn by David Burger and Steven Nys in 1985 for the Historic American Buildings Survey.

4. H. Wayne Price and Keith A. Sculle, "The 'Doughnut' and 'Oval' Barns of Ogle and Stephenson Counties, Illinois: An Architectural Survey," *Pioneer America Society Transactions* 9 (1986): 31-38.

BIBLIOGRAPHY

The American Agriculturist 1-52 (1843-93).

Arthur, Eric and Dudley Whitney. *The Barn, a Vanishing Landmark in North America.* Greenwich, Conn.: New York Graphic Society, 1972.

Bealer, Alex W. *The Log Cabin: Homes of the North American Wilderness.* Barre, Mass.: Barre, 1978.

Brakebill, David. "Folk Architecture: The Cantilevered Barns of Eastern Tennessee." B.Arch. thesis, Univ. of Tennessee, 1973. In Norbert F. Riedl Papers. Box 1: Architecture, Folder 4.

Breazeale, J. W. M. *Life as It Is; or Matters and Things in General.* Knoxville: James Williams, 1842.

Bokum, Hermann. *The Tennessee Hand-Book and Immigrant's Guide.* Philadelphia: J. B. Lippencott, 1868.

Buckwalter, Donald W. "Effects of Early Nineteenth Century Transportation Disadvantage on the Agriculture of Eastern Tennessee." *Southeastern Geographer* 27 (1987): 18-37.

Bugge, Gunnar, and Christian Norberg-Schulz. *Stav og Laft i Norge* [Early Wooden Architecture in Norway]. Oslo: Norske Arkitekters Landsforbunds, 1969.

Burns, Inez E. *History of Blount County, Tennessee: From War Trail to Landing Strip, 1795-1955.* Maryville, Tenn.: Mary Blount Chapter Daughters of the American Revolution and Tennessee Historical Commission, 1957.

——— "Settlement and Early History of the Coves of Blount County, Tennessee." *East Tennessee Historical Society Publications* 24 (1952): 44-67.

Buxton, David. *The Wooden Churches of Eastern Europe: An Introductory Survey.* Cambridge: Cambridge Univ. Press, 1981.

Campbell, John C. *The Southern Highlander and His Homeland.* New York: Russell Sage Foundation, 1921. Rpt. Lexington: Univ. Press of Kentucky, 1969.

Case, Earl C. *The Valley of East Tennessee: The Adjustment of Industry to Natural Environment.* Nashville: State Geological Survey, 1925.

Crawford, W. *Tennessee: The Land Where God Has Set His Seal of Love and Nature's Garden Home Is Built.* Nashville: James T. Camp, 1899.

Creekmore, Pollyanna. "Matthew Tarwater Family Bible." *Smoky Mountain Historical Society Newsletter* 14, no. 4 (Winter 1986): 114-15.

———, and Blanche C. McMahon, comps. "Population Schedule of the United States Census of 1850 (7th Census) for Sevier County, Tennessee." Typescript. Knoxville, 1953.

DeArmond, Nora Harbin. "Bits and Pieces of History about Smoky Mountains Foothills People: Ingle Families." *Smoky Mountain Historical Society Newsletter* 16, no. 2 (Winter 1990): 98-100.

Dockter, Albert W., Jr. "Zwingli-ites in Sevier and Knox Counties?????"

Smoky Mountain Historical Society Newsletter 15, no. 1 (Spring 1989): 10-13.

Duggan, W. L. "Facts about Sevier County." Typescript. Sevierville, Tenn., 1910.

Durand, Loyal, Jr., and Elsie Taylor Bird. "The Burley Tobacco Region of the Mountain South." *Economic Geography* 26 (1950): 274-300.

Egerton, John. *Generations: An American Family.* New York: Simon and Schuster, 1986.

Eller, Ronald D. "Land and Family: An Historical View of Preindustrial Appalachia." *Appalachian Journal* 6, no. 2 (Winter 1979): 83-109.

Ensminger, Robert F. "A Search for the Origin of the Pennsylvania Barn." *Pennsylvania Folklife* 30, no. 2 (Winter 1980-81): 50-70.

Evans, E. Raymond. "Fort Marr Blockhouse, Last Evidence of America's First Concentration Camp." *Journal of Cherokee Studies* 2 (Spring 1977): 256-63.

Featherstonhaugh, George William. *Excursion through the Slave States, from Washington on the Potomac, to the frontier of Mexico; with sketches of popular manners and geological notices.* London: J. Murray, 1844.

Fink, Daniel. *Barns of the Genesee County, 1790-1915.* Geneseo N.Y.: James Brunner, 1988.

Fischer, David Hackett. *Albion's Seed: Four British Folkways in America.* New York: Oxford Univ. Press, 1989.

Fleischhauer, Carl, and Howard W. Marshall, eds. *Sketches of South Georgia Folklife.* Washington, D.C.: Library of Congress, 1977.

Folmsbee, Stanley J., Robert E. Corlew, and Enoch L. Mitchell. *Tennessee: A Short History.* Knoxville: Univ. of Tennessee Press, 1969.

Fowler, Aileen Whaley. "Fun, Friends and Fellowship: Recreation and Amusements," in *The Gentle Winds of Change: A History of Sevier County, Tennessee, 1900-1930.* Maryville, Tenn.: Smoky Mountain Historical Society, 1986: 179-98.

Garrison, J. Ritchie, Bernard L. Herman, and Barbara McLean Ward, eds. *After Ratification: Material Life in Delaware, 1789-1820.* Newark: Museum Studies Program, Univ. of Delaware, 1988.

Geschwend, Max. *Schweizer Baurenhäuser: Material, Konstruktion und Einteilung.* Bern: Verlag Paul Haupt, 1983.

Glass, Joseph W. *The Pennsylvania Culture Region: A View from the Barn.* Ann Arbor: UMI Research Press, 1986.

Glassie, Henry. "The Appalachian Log Cabin." *Mountain Life and Work* 39, no. 4 (Winter 1963): 5-14.

———. "The Double Crib Barn in South Central Pennsylvania." *Pioneer America* 1, no. 1 (1969): 9-16; 1, no. 2 (1969): 40-45; 2, no. 1 (1970): 47-52; 2, no. 2 (1970): 23-34.

———. *Folk Housing in Middle Virginia.* Knoxville: Univ. of Tennessee Press, 1975.

———. "The Old Barns of Appalachia." *Mountain Life and Work* 40, no. 2 (Summer 1965): 21-30.

———. *Pattern in the Material Folk Culture of the Eastern United States.* Philadelphia: Univ. of Pennsylvania Press, 1968.

———. "The Pennsylvania Barn in the South." *Pennsylvania Folklife* 15, no. 2 (1966): 8-19.

———. "The Pennsylvania Barn in the South." *Pennsylvania Folklife* 15, no. 4 (1966): 12-25.

———. "The Variation of Concepts within Tradition: Barn Building in Otsego County, New York." in *Man and Cultural Heritage.* Vol. 5 of *Geoscience and Man*, edited by H. J. Walker and W. G. Haag. Baton Rouge: Louisiana State Univ., 1974, 177-235.

Gray, Lewis Cecil. *History of Agriculture in the Southern United States to 1860.* Washington, D. C.: Carnegie Institution, 1933.

Greene, Elmer A., comp. "Population Schedule of the United States Census of 1860 (8th U.S. Census) for Sevier County, Tennessee." Typescript. Kingsport, Tenn., 1971.

Halsted, Byron D. *Barn Plans and Outbuildings.* New York: Orange Judd, 1881.

Heikkenen, Herman J., and Mark R. Edwards. "The Key-Year Dendrochronology Technique and Its Application in Dating Historic Structures in Maryland." *Association for Preservation Technology Bulletin* 15, no. 3 (1983): 2-25.

Herman, Bernard L. *Architecture and Rural Life in Central Delaware, 1700-1900.* Knoxville: Univ. of Tennessee Press, 1987.

Holan, Jerri. *Norwegian Wood: A Tradition of Building.* New York: Rizzoli, 1990.

Hubka, Thomas C. *Big House, Little House, Back House, Barn: The Connected Farm Buildings of New England.* Hanover, N.H.: Univ. Press of New England, 1984.

Hutslar, Donald A. "The Ohio Farmstead: Farm Buildings as Cultural Artifacts." *Ohio History* 90, no. 3 (Summer 1981): 221-37.

Johnson, Clifton. *Highways and Byways of the South.* New York: Macmillan, 1904.

Jordan, Terry G. "Alpine, Alemannic, and American Log Architecture." *Annals of the Association of American Geographers* 70 , no. 2 (1980): 154-80.

———. *American Log Buildings: An Old World Heritage.* Chapel Hill: Univ. of North Carolina Press, 1985.

———. "A Forebay Bank Barn in Texas." *Pennsylvania Folklife* 30, no. 2 (Winter 1980-81): 72-77.

———. "Moravian, Schwenkfelder, and American Log Construction." *Pennsylvania Folklife* 33, no. 3 (Spring 1984): 98-124.

———. "Some Neglected Swiss Literature on the Forebay Bank Barn," *Pennsylvania Folklife* 37, no. 2 (1987-88), 75-80.

———. *Texas Log Buildings: A Folk Architecture.* Austin: Univ. of Texas Press, 1978.

———, and Matti Kaups. *The American Backwoods Frontier: An Ethnic and Ecological Interpretation.* Baltimore: Johns Hopkins Univ. Press, 1989.

Kephart, Horace. *Our Southern Highlanders.* New York: Macmillan, 1922.

Killebrew, Joseph Buckner. *Introduction to the Resources of Tennessee.* Nashville: Tavel, Eastman & Howell, 1874.

———. *Report on the Culture and Curing of Tobacco in the United States.* Washington, D.C.: Government Printing Office, 1884.

———. *Tennessee: Its Agricultural Resources and Mineral Wealth.* Nashville: Tavel, Eastman & Howell, 1874.

———. *Tobacco: Its Culture in Tennessee.* Nashville: Tavel, Eastman & Howell, 1876.

King, Veta. "Dupont's History Started at Line Where Dew Won't Form." *Knoxville News-Sentinel South,* Mar. 20, 1991.

Klöckner, Karl. *Der Blockbau: Massivbauweise in Holz.* Munich: Callwey Verlag, 1982.

Kniffen, Fred B. "Folk Housing: Key to Diffusion." *Annals of the Association of American Geographers* 55, no. 4 (1965): 549-77.

———. "On Corner-Timbering." *Pioneer America* 1, no. 1 (Jan. 1969): 1-8.

———, and Henry Glassie. "Building in Wood in the Eastern United States: A Time-Place Perspective." *Geographical Review* 56 (1966): 40-66.

Kofoid, Charles A. *Termites and Termite Control.* Berkeley: Univ. of California Press, 1934.

Kollmorgen, Walter M. "Observations on Cultural Islands in Terms of Tennessee Agriculture." *East Tennessee Historical Society's Publications* 16 (1944): 65-78.

Letchner, Peter. "The Breaks, Virginia." *Pioneer America* 4, no. 2 (July 1972): 1-7.

Linn, Beulah Duggan. "50 Year History of Educational Growth." *Sevier County News-Record,* Jan. 15, 1976.

———. "Henry Butler of Sevier County, Tennessee." *Smoky Mountain Historical Society Newsletter* 8, no. 1 (Spring 1982): 2-3.

———. "A History of the Bank of Sevierville." Sevierville, 1988.

———. "Joshua Tipton, Forgotten Revolutionary Soldier." *Sevier County News-Record,* July 22, 1976.

Little, Edith B. *Blount County Tennessee Cemetery Records.* Evansville, Ind.: Whipporwill Publications, [1981].

Marshall, Howard Wight. "The Pelster Housebarn: Endurance of Germanic Architecture on the Midwestern Frontier." *Material Culture* 18, no. 2 (Summer 1986): 65-104.

Maryville College Political and Social Science Dept. *Social Survey of Blount County, 1930.* Maryville, Tenn., 1930.

Michaux, François André. *Travels to the Westward of the Allegany Mountains in the States of Ohio, Kentucky and Tennessea.* London: B. Crosby, 1805.

Moffett, Marian, and Lawrence Wodehouse. *The Cantilever Barn in East Tennessee.* Knoxville: Univ. of Tennessee School of Architecture, 1984.

———. "The Cantilever Barn in East Tennessee." *Pioneer America Society Transactions* 9 (1986): 17-22.

———. "Cantilever Barns of East Tennessee." *The Tennessee Conservationist* 52, no. 1 (Jan.-Feb. 1986): 2-3.

Montell, William Lynwood and Michael Lynn Morse. *Kentucky Folk Architecture.* Lexington: Univ. Press of Kentucky, 1976.

Moore, Harry B. *Wood-Inhabiting Insects in Houses: Their Identification, Biology, Prevention and Control.* Washington, D.C.: Dept. of Housing and Urban Development, 1979.

Morgan, John T. "The Decline of Log House Construction in Blount County, Tennessee." Doctoral dissertation, Univ. of Tennessee, Knoxville, 1986.

———. *The Log House in East Tennessee.* Knoxville: Univ. of Tennessee Press, 1990.

———, and Ashby Lynch, Jr. "The Log Barns of Blount County, Tennessee." *Tennessee Anthropologist* 9, no. 2 (1984): 85-103.

Morris, Eastin. *The Tennessee Gazetteer or Topographical Dictionary.* Nashville: W. Hasell, Hunt, 1834.

Naismith, Robert J. *Buildings of the Scottish Countryside.* London: Victor Gollancz Ltd., 1985.

Newton, Milton B., Jr., and Linda Pulliam-DiNapoli. "Log Houses as Public Occasions: A Historic Theory." *Annals of the Association of American Geographers* 67 (1977): 360-83.

Noble, Allen G. *Wood, Brick, and Stone: The North American Settlement Landscape. Vol. 2: Barns and Farm Structures.* Amherst: Univ. of Massachusetts Press, 1984.

———, ed. *To Build in a New Land: Ethnic Landscapes in North America.* Baltimore: Johns Hopkins Univ. Press, 1992.

O'Neil, Nancy L. "The Swaggerty Blockhouse." *Smoky Mountain Historical Society Newsletter.* 11, no. 1 (Spring 1985): 4-7.

Opolovnikov, Alexander, and Yelena Opolovnikova. *The Wooden Architecture of Russia: Houses, Fortifications, Churches.* New York: Harry N. Abrams, 1989.

Paine, Thomas Harden. *Short Sketches of the Counties of Tennessee.* Nashville: Printed for the Secretary of Agriculture, 1901.

Periam, Jonathan. *The Home and Farm Manual.* 1884. Rpt. New York: Greenwich House, 1984.

Petersen, Eugen Adolf Hermann, and Felix von Luschan. *Reisen im Südwestlichen Kleinasien.* Vienna, 1899.

Phleps, Hermann. *The Craft of Log Building: A Handbook of Craftsmanship in Wood.* 1942. Rpt. Ottawa: Lee Valley Tools, Ltd., 1982.

Price, H. Wayne, and Keith A. Sculle. "The 'Doughnut' and 'Oval' Barns of Ogle and Stephenson Counties, Illinois: An Architectural Survey." *Pioneer America Society Transactions* 9 (1986): 31-38.

Rambo, Beverly Nelson. *The Rambo Family Tree: Descendants of Peter Gunnarson Rambo, 1611-1986.* Decorah Iowa: Anundsen, 1986.

Rawson, Richard. *Old Barn Plans.* New York: Mayflower Books, 1979.

Rehder, John B. "The Scotch-Irish and English in Appalachia." In To Build in a New Land, edited by Allen G. Noble. Baltimore: Johns Hopkins Univ. Press, 1992: 95-118.

The Republican Star. Sevierville, Tenn. Oct. 1891-Mar. 1892.

Riedl, Norbert F., Papers. MS 954, Box 1: Architecture, Folders 3 and 4. Special Collections Library of the Univ. of Tennessee, Knoxville.

Roberts, Warren B. "Folk Architecture in Context: The Folk Museum." *Proceedings of the Pioneer America Society* 1 (1972): 34-50.

———. "The Tools Used in Building Log Houses in Indiana." *Pioneer America* 9, no. 1 (1977): 32-61.

Rooney, John F., Wilbur Zelinsky, and Dean R. Louder, eds. *This Remarkable Continent: An Atlas of United States and Canadian Society and Cultures.* College Station: Texas A&M Univ. Press, 1982.

Roth, Margaret Ann. "The End of Isolation: Transportation and Communication," in *The Gentle Winds of Change: A History of Sevier County 1900-1930.* Maryville, Tenn.: Smoky Mountain Historical Society, 1986: 54-79.

Rudenko, Sergei Ivanovich. *Frozen Tombs of Siberia: The Pazyryk Burials of Iron Age Horsemen.* Trans. M. W. Thompson. Berkeley: Univ. of California Press, 1970.

Schafer, Joseph. *The Social History of American Agriculture.* New York: Macmillan, 1936.

Schultz, LeRoy G. *Barns, Stables, and Outbuildings: A World Bibliography in English, 1700-1983.* Jefferson, N.C.: McFarland, 1986.

Scoates, Daniels. *Farm Buildings.* College Station: A&M College of Texas, 1937.

Sevier County Bicentennial Committee. "Sevier County Saga." Typescript and photocopies of newspaper articles. Sevierville, Tenn., 1976.

Sevier County Historical Society. Sevier County Cemeteries. n.d.

"Sevier County Members of Tennessee General Assembly." *Sevier County News-Record,* July 1, 1976.

Sloane, Eric *An Age of Barns.* New York: Funk & Wagnalls, 1967.

———. *American Barns and Covered Bridges.* New York: Wilfred Funk, 1954.

———. *A Museum of Early American Tools.* New York: Funk & Wagnalls, 1964.

Smith, J. Gracy. *A Brief Historical, Statistical, and Descriptive Review of East Tennessee, U.S.A.* London: J. Leath, 1842.

Smoky Mountain Historical Society. *The Gentle Winds of Change: A History of Sevier County, Tennessee, 1900-1930.* Maryville, Tenn.: Smoky Mountain Historical Society, 1986.

——— *In the Shadow of the Smokies: Sevier County Cemeteries.* Sevierville: Smoky Mountain Historical Society, 1984.

Soike, Lowell J. *Without Right Angles: The Round Barns of Iowa.* Des Moines: Iowa State Office of Historic Preservation, 1983.

Stilgoe, John R. *Common Landscape of America, 1580 to 1845.* Cambridge: Harvard Univ. Press, 1982.

Strother, David Hunter. *The Old South Illustrated, by Ponte Crayon.* Edited by Cecil D. Eby, Jr. Chapel Hill: Univ. of North Carolina Press, 1959.

Sudworth, George B., and Joseph Buckner Killebrew. *The Forests of Tennessee: Their Extent, Character, and Distribution.* Nashville: M.E. Church, South, 1897.

Templin, David, and Elaine Wells, comps. "Population Schedule of the U.S. Census of 1840 for Sevier County, Tennessee." Typescript. Maryville, Tenn., 1981.

Tennessee. Bureau of Agriculture. *Bulletin of Farms, Residences, Etc. for Sale in Tennessee, Collected, Registered, and Advertised by A. J. McWhirter.* Nashville: State of Tennessee, 1883.

Trentham, Joan T., comp. "Population Schedule of the U.S. Ninth Census for Sevier County, Tennessee, 1870." Typescript Sevierville, Tenn., 1980.

Trotter, Isaac (1810-85). Papers. MS 230. Special Collections Library of the Univ. of Tennessee, Knoxville.

United States Bureau of the Census. *Eighth Census of Agriculture* (1860). Microfilm.

———. *Eighth Census of Population* (1860). Microfilm.

———. *Fifth Census of Population* (1830). Microfilm.

———. *Ninth Census of Agriculture* (1870). Microfilm.

———. *Ninth Census of Population* (1870). Microfilm.

———. *Seventh Census of Agriculture* (1850).

———. *Seventh Census of Population* (1850). Microfilm.

———. *Sixth Census of Population* (1840). Microfilm.

———. *Tenth Census of Agriculture* (1880). Microfilm.

———. *Tenth Census of Population* (1880). Microfilm.

———. *Thirteenth Census of Population* (1910). Microfilm.

———. *Twelfth Census of Population* (1900). Microfilm.

Upton, Dell. *Holy Things and Profane: Anglican Parish Churches in Colonial Virginia.* New York: Architectural History Foundation and Cambridge: MIT Press, 1986.

Virginia, Tennessee and Georgia Air Line. *The Scenic Attractions and Summer Resorts along the Railways of the Virginia, Tennessee, and Georgia Air Line, the Shenandoah Valley Railroad, the Norfolk and Western Railroad, and the East Tennessee, Virginia and Georgia Railroad. . . .* New York: Aldine Press, 1883.

Warner, Charles Dudley. *On Horseback: A Tour in Virginia, North Carolina, and Tennessee.* Boston: Houghton Mifflin, 1888.

Weslager, C. A. *The Log Cabin in America: From Pioneer Days to the Present.* New Brunswick: Rutgers Univ. Press, 1969.

West, Carroll Van. *Tennessee Agriculture: A Century Farm Perspective.* Nashville: Tennessee Dept. of Agriculture, 1986.

West Virginia Univ. Library. *Appalachian Bibliography.* Morganton: West Virginia State Library, 1980.

Wheeler, William Bruce. "Government and Politics" in *The Gentle Winds of Change: A History of Sevier County, 1900-1930.* Maryville, Tenn.: Smoky Mountain Historical Society, 1986: 218-36.

Wiggenton, Eliot, ed. *The Foxfire Book.* New York: Doubleday, 1972.

INDEX

Page references in **bold** type indicate illustrative material: photographs, maps, drawings, or charts.

agricultural census, 30, 130 n1
Agricultural Extension Service, 51, 75
agricultural profile of barn builders, 29–34, **80–83**
agriculture in East Tennessee, 29
American Revolution, 28
Anatolia, 23, **25**
Andes, John (Sevier 32), **15**, 32, **33**, 35, **36**, **43**, **55**, **82**, 128 n9
Andes, John W. (Sevier 3), **15**, 32, **33**, **34**, 128 n10
Andes, Riley H. (Sevier 18), **15**, 32, **33**, **34**, 128 n10
Appalachian log cabin, 18
Atchley, Jesse (Sevier 151), 126 n11

bank barns, 17, 21–23
bank barns in Knox County, 22–23
Barger barn (Botetourt County Virginia), 3, 25, **26**, 125 n2
blockhouses, 17, 23, **24–25**, 53, 127 n27
Bohannon, James (Sevier 60), **15**, 45, 46, 129 n23, 129 n24
Boyer barn, Cocke County, 22, **23**
Buckwalter, Donald W., 50, 130 n4

burley tobacco, 40, 49
Burrison, John, 125 n4
Butler, Henry (Sevier 156), **15**, **59**, **69**, 71, 132 n25
Butler, Horatio (Sevier 136), **15**, 35, **36**, **55**, **59**, **61**, 63, **64**, 65, 71, 130 n8
Butler, James H. (Sevier 180), **15**, **59**, **61**, 65–66

Cades Cove, xiii, **xiv**, 8, **9**, 76
Campbell, John C., 126 n14, 126 n10, 127 n3
Case, Earl C., 129 n1
Catlett, William (Sevier 37), **15**, **33**, **34**, 35, **36**, **55**, 128 n13
Caudill barn (Kentucky), 25, **26**, 125 n3
central Europe, 19
Cherokee Indians, 27–29
climate of East Tennessee, 48
Corle barn (Pennsylvania), 24–25
Corlew, Robert E., 127 n2
corner men, 16
corner timbering on barns, 4, 125 n7
Creekmore, Pollyanna, 129 n19
cultural diffusion, 17, 18

dates of barn construction, 13, **14**, **15**, 16
DeArmond, Nora Harbin, 129 n22
Delaware settlements, 18, 72
dendrochronology, 13, 125 n10
Dockter, Albert W., Jr., 129 n25
Duggan, Pryor (Sevier 7), **15**, 31, **81**, 128 n7

early settlement in East Tennessee, 27–29
Egerton, John, 127 n3
English barns, 17
English settlers, 27, 29
Ensminger, Robert, 20, 24, 127 n14, 127 n15
Evans, Raymond, 127 n 25

fachwerk, 19
families of barn builders, 40–44
Fischer, David H., 127 n3
Folmsbee, Stanley J., 127 n2
Fort Loudoun, 27
Fort Marr, Polk County, 23, **24**
foundations of barns, 3–4
four-crib cantilever barns, 12
Fowler, Aileen Whaley, 131 n17

French and Indian War, 28
French settlers, 29
Frontier Culture Museum of Virginia, 25, **26**

German barn types, 20–23
German settlers, 27, 28
Geschwend, Max, 126 n12
Glassie, Henry, 1, **2**, 17, 18, 21, 125 n7, 125 n9, 126 n7, 127 n17, 127 n20
Great Smoky Mountains National Park, xiii, 76
Greene County barns, 37, 39, 65
Groton, Betsy, 126 n10

half-double cantilever barns, 11
Halsted, Byron D., 126 n 13
Hamilton, Gary, 23, 127 n22
Henderson, Robert (Sevier 138), **15**, **59**, **67**, **69**, 70–71, 131 n19
Herman, Bernard, 126 n4
Householder, Mattie Belle Butler, 71–72
Hubka, Thomas C., 132 n2

illiteracy, 46, 129 n26
Ingle barns (Sevier 29, 89, 113, 114), **xv**, 5, 7, **15**, 40, **44**, **45**

Irish settlers, 27
isolation, 49–52

Johnson County barns, **50**
Jordan, Terry, 18, 20, 23, 72, 126 n4, 126 n5, 126 n8, 127 n16, 127 n28

Karelia, 19
Kaups, Matti, 18, 72, 126 n4
Killebrew, Joseph Buckner, 18, 30, 49, 125 n8, 126 n9, 128 n4, 128 n6, 130 n3, 130 n5
King, Veta, 128 n17
Kniffen, Fred B., 21, 125 n7, 127 n19
Knox County bank barns, 22–23, 127 n23
Kollmorgen, Walter M., 127 n21

landscape and agriculture, 47–49
Linn, Beulah Duggan, 71, 128 n12, 128 n18, 129 n23, 129 n24, 129 n26, 130 n2, 130 n5, 130 n6, 130 n8, 131 n11, 131 n13, 131 n16, 132 n26, 132 n27, 132 n29
Little, Edith B., 128 n5
local groupings of similar barns, 37–40
location of barns, xv, **xvi**, **2**
loft construction, 6, **7**
log construction, 17–19
Luschan, Felix von, 127 n27
Lynch, Ashby, 3, 125 n5

McCampbell, James H. (Blount 9), 12–**13**, **14**
McMahan, George Washington (Sevier 2), **15**, **34**–35, **67**, 128 n14, 128 n15

McMahan, Thomas DeArnold Wilson (Sevier 10 and 14), 8–9, **10**, **15**, **48**, **61**, 62
Marshall, John (Sevier 61), **15**, 31, **80**
Marshall, Robert (Sevier 137), **15**, 31, **59**, **61**, **64**, 65, 70, **80**, 131 n15
Martin, Charles, 26, 125 n3
Michaux, François André, 127 n1
Middle Creek (Sevier County), historical and agricultural profile, 53–56
Mitchell, Enoch L., 127 n2
Montell, William Lynwood, 125 n3
Moore, Harry B., 129 n 2
Morgan, John, 3, 18, 125 n5, 125 n7, 126 n11
Morse, Michael Lynn, 125 n3
Museum of Appalachia, 76

Naismith, Robert J., 126 n1
Newton, Milton B., Jr., 126 n12
Noble, Alan, 21, **22**, 76, **77**, 127 n18

O'Neil, Nancy L., 127 n26
Ogle, John (Sevier 172), **15**, **59**, **61**, **64**, 71, 132 n23
Opolovnikov, Alexander, 127 n24
Opolovnikova, Yelena, 127 n24

Pennsylvania German barns, 20–23
Pennsylvania Germans, 17, 18, 22–23, 127 n21
Petersen, Eugen Adolf Hermann, 127 n27
Phleps, Hermann, 19, 126 n2, 127 n13
precedent for cantilever design, 19–20

Price, H. Wayne, 132 n4
Pulliam-DiNapoli, Linda, 126 n12

railroads in East Tennessee, 50–51
Rambo, Beverly Nelson, 132 n29
Rambo, Robert Myomey, 72
Reagan, Richard (Sevier 52), **15**, 35, **36**, 37, 46, **55**, **82**, 128 n16
regional nature of barn designs, 75, 76
Rehder, John B., 126 n1
Republican Star, The, 51
river transportation, 51
Roberts, Stephen H. (Sevier 166), **15**, **59**, **67**, 71
Roberts, Warren, 126 n6
Roberts, William M. (Sevier 165), **15**, **59**, **67**, 71, 131 n21
Roberts, William Phillip (Sevier 173), **15**, 35, **36**, **55**, **59**, **67**–68, **69**, 70, 71, 72, 131 n16
Robertson, Dio Cleason (Sevier 28), 12, **15**, **59**, **67**, **69**, 71, 131 n20
roofs of barns, 7
Roth, Margaret Ann, 130 n7
Rudenko, Sergei Ivanovich, 126 n3
Russia, 17, **19**, 23, **24**

Scandinavia, 17, 18, 19
Schultz, LeRoy G., 1
Scots-Irish settlers, 21, 27, 28, 29, 47
Sculle, Keith A., 132 n4
Seaton, Jacob H. (Sevier 152), **15**, **59**, **67**, **69**, 71, 132 n24
Seaton, James H. (Sevier 23), **15**, **59**, **67**, 68, **69**, **70**, 71, 131 n17
Seaton, John A. (Sevier 21), **15**, **59**, **67**, 68, **69**, 70–71, 131 n18
Sharp, John (Sevier 74), 9, 10
Sharp barns, Cedar Bluff (Sevier 11, 12, 15), **12**, **13**, 37

Shenondoah Valley, 21
Shields, Robert, 53–54
Shields Fort, 53–54, 60, 130 n6
siding on barns, 7
single-crib doublecantilever barns, 12
Smoky Mountain Historical Society, 73, 128 n5, 130 n3, 132 n1
Soike, Lowell J., 132 n3
soil in East Tennessee, 48
Starr, Nelle, 39
State of Franklin, 28
Strother, David Hunter, 130 n6
Suttles barn (Sevier 170), 16
Sutton, William M. (Sevier 139), **15**, **59**, **67**, **69**, 72, 132 n28
Swaggerty blockhouse, Cocke County, 23, **25**, 127 n26
Swiss forebay barns, 17, **20**, 37

Tarwater, James Rogers (Sevier 168), **15**, 40, **41**, **43**, **83**, 129 n19
Tarwater, Levi, (Sevier 85), **15**, 40, **41**, **43**
Tarwater, Matthew (Sevier 9), **15**, 40, **41**, **43**, **83**, 128 n19
Tarwater, William Dowell (Sevier 83), **15**, 40, **41**, **43**, **83**, 129 n19
Tennessee Valley Authority (TVA), 74–75
termites, 48–49, 129 n2
Territory South of the Ohio, 28
threshing floor, 5
Tipton homeplace barn, Cades Cove (Blount 8), xiii, **xiv**, 8, **9**, 76
tobacco barns, 49, 65
transportation in East Tennessee, 49–51
treaties with the Indians, 28
Trotter, Amos C. F. (Sevier 19), **15**, **59**, **61**, **64**, 65, **66**, 131 n10

Trotter, Isaac (Sevier 20), **15**, **59**, **61**, 63, **64**, 65, 130 n3, 130 n10
Trotter, John (Sevier 5), **15**, 35, **36**, **55**, **59**, 60, **61**, 62, **63**, **64**, 67, 71, 130 n7
Trotter, William Harrison (Sevier 6), **15**, 35, **36**, **55**, **59**, **61**, 63, **64**, 65, 130 n9, 131 n13, 131 n14

Trotter, William J., 71, 132 n26
two-crib double-cantilever barns, 8–10, 21, 35
two-crib single cantilever barns, 10–11
types of cantilever barns, 8–13

Umgebindehaus, 21
use of cribs, 4, 73, 75

Warner, Charles Dudley, 125 n8, 130 n6
Washington County barns, 49, 65
Webb, William W. (Sevier 22), **15**, 31, **59**, **67**, **69**, 71, **81**, 131 n22
Wells, Elaine, 130 n5
Weslager, C. A., 18, 126 n4
West, Carroll Van, 129 n20

Wheeler, William Bruce, 131 n12
White, Meddy, 60, **61**
Whitehead, John W. (Sevier 135), 45–46, 129 n25
Wiggenton, Eliot, 126 n10
wood used in barn construction, 5
Woods, Frank, 126 n10